UNNECESSARY EXPENSE

CHARLES THEUER M.D., Ph.D.
WITH BONNE ADAMS, MARK WIGGINS, AND SCOTT BROWN

UNNECESSARY EXPENSE

An **ANTIDOTE** *for the*
BILLION-DOLLAR DRUG PROBLEM

ForbesBooks

Copyright © 2021 by Charles Theuer. M.D., Ph.D.

All rights reserved. No part of this book may be used or reproduced in any manner whatsoever without prior written consent of the author, except as provided by the United States of America copyright law.

Published by ForbesBooks, Charleston, South Carolina.
Member of Advantage Media Group.

ForbesBooks is a registered trademark, and the ForbesBooks colophon is a trademark of Forbes Media, LLC.

Printed in the United States of America.

10 9 8 7 6 5 4 3 2 1

ISBN: 978-1-95086-357-0
LCCN: 2021911886

Cover design by Carly Blake.
Layout design by Mary Hamilton.

This custom publication is intended to provide accurate information and the opinions of the author in regard to the subject matter covered. It is sold with the understanding that the publisher, Advantage|ForbesBooks, is not engaged in rendering legal, financial, or professional services of any kind. If legal advice or other expert assistance is required, the reader is advised to seek the services of a competent professional.

Advantage Media Group is proud to be a part of the Tree Neutral® program. Tree Neutral offsets the number of trees consumed in the production and printing of this book by taking proactive steps such as planting trees in direct proportion to the number of trees used to print books. To learn more about Tree Neutral, please visit **www.treeneutral.com**.

Since 1917, Forbes has remained steadfast in its mission to serve as the defining voice of entrepreneurial capitalism. ForbesBooks, launched in 2016 through a partnership with Advantage Media Group, furthers that aim by helping business and thought leaders bring their stories, passion, and knowledge to the forefront in custom books. Opinions expressed by ForbesBooks authors are their own. To be considered for publication, please visit **www.forbesbooks.com**.

This book is dedicated to cancer patients and the health care providers who dedicate their lives to treating them.

CONTENTS

FOREWORD ix

INTRODUCTION 1

CHAPTER ONE 7
THE BILLION-DOLLAR DRUG PROBLEM

CHAPTER TWO 21
JOURNEY FROM ONCOLOGY SURGEON
TO BIOTECHNOLOGY EXECUTIVE

CHAPTER THREE 35
BIOTECHNOLOGY LESSONS

CHAPTER FOUR 51
BIG PHARMA LESSONS

CHAPTER FIVE 69
CROS—THE PITFALLS OF BUSINESS
AS USUAL IN DRUG DEVELOPMENT

CHAPTER SIX **87**
UNDERSTANDING AND
NAVIGATING THE FDA

CHAPTER SEVEN **97**
THE TRACON PRODUCT
DEVELOPMENT PLATFORM

CHAPTER EIGHT **117**
HARNESSING GLOBAL INNOVATION
FOR US PATIENTS

CHAPTER NINE **137**
INTEGRATING TECHNOLOGY AS
PART OF THE TRACON PDP

CHAPTER TEN **151**
THE FINER POINTS OF COMMERCIALIZING
AN ONCOLOGY DRUG

SUMMARY **157**

ACKNOWLEDGMENTS **161**

GLOSSARY **163**

ABOUT THE LEAD AUTHOR **175**

OUR SERVICES **177**

FOREWORD

The Future of Clinical Trials

Over the last several decades, we have witnessed a public debate about the rising cost of discovering and developing new drugs. Currently, the cost of discovering, developing, and commercializing a drug varies, but most pharmaceutical and biotechnology professionals agree that it exceeds $1 billion to bring a drug to market, with the average duration of the clinical trials needed to approve a new drug being more than ten years, depending on the therapeutic area. Wouters and colleagues concluded that cancer drugs are among the most expensive to develop, with a median estimate of $2.8 billion.[1] More recently, Pfizer stated the research and development costs of its SARS Cov-2 (COVID-19) vaccine approached $1 billion. Despite the fact that the development of the COVID-19 vaccine was one of the most efficient vaccine research collaborations between for-profit and nonprofit institutions that we have observed, with governments, academia, and pharmaceutical companies working in an unprecedented level of synchronicity to accelerate vaccine discovery, the cost of research and development remained staggering.

---

1   O. J. Wouters, M. McKee, and J. Luyten, "Estimated Research and Development Investment Needed to Bring a New Medicine to Market, 2009-2018," *JAMA* 323: 834-43, 2020.

As our understanding of diseases and their causes improves, patients and caregivers continue to demand safer and more efficacious drugs. There is increased pressure to consider the genetic heterogeneity of patient populations to allow for the development of drugs that are highly selective and have fewer side effects. Developing safe and efficacious drugs requires not only detailed and thorough research but also large clinical trials to thoroughly understand drug effects before companies can approach the FDA for approval. All of these research and regulatory requirements add to the cost of developing a new drug.

Researchers generally agree that the conventional drug discovery process is costly and time-consuming with very high attrition rates and very low success rates. How do we approach the disparity between the high cost of drug discovery and vast unmet medical needs? Over the years, our understanding of the genetic basis of disease has increased, and the costs of conducting genetic analyses to identify certain diseases has decreased considerably. The application of computer modelling to understand the interactions between proteins and drug candidates has not only accelerated the discovery of new therapeutics, but has also improved precision targeting of drugs for particular diseases. Recent application of artificial intelligence has made significant inroads into the field of drug discovery. This is mainly due to the increased availability of large databases that can be leveraged to create more accurate machine learning models, thereby lowering the barrier to entry for researchers. Currently, artificial intelligence is being employed to identify and repurpose old or existing drugs for new therapies, a practice which also reduces the cost of research and development. In recent years several biotechnology companies have developed low-cost diagnostic kits to rapidly identify diseases based on specific biomarkers. The cost of these test kits has decreased, making it quite feasible to diagnose a large number of patients in a short period of time.

FOREWORD

These advances across multiple areas of drug discovery should have helped lower the cost of discovering and advancing drug candidates. However, despite these myriad advances, the overall cost of research and development of drugs continues to increase. If the actual procedural costs of research and development have decreased, why have overall costs continued to climb? A close examination of the breakdown of these costs reveals that approximately 70 percent of the cost of developing a drug is attributed to conducting clinical trials (Phase 1, 2, and 3)—in other words, despite the improvements in drug discovery technologies, the cost of drug development continues to climb steeply because of the rate of increase in clinical trial costs. As such, it is extremely rare for small and midsized biotechnology companies to conduct clinical trials on their own due to lack of funding and internal expertise. Consequently, these companies usually partner with larger companies that have deep expertise in a given therapeutic area and have the resources to staff an internal clinical development team. Apart from partnering with larger companies, it is also increasingly common for many companies to work with clinical contract research organization (CROs) to conduct clinical trials. While outsourcing to a clinical CRO may appear to be a cost-efficient solution, the practice suffers from lack of ownership from the CROs. CROs usually operate on a fee-for-service reimbursement model, and once they deliver the service, they move on to next customer—their incentives are not aligned with those of the drug developers. As such, we need a new business model where the partners responsible for clinical trial implementation are as motivated as the companies who discovered the drug candidate. It is important for a company to work with a partner who has deep-seated knowledge in a given therapeutic area and who is equally vested in the success of a new therapeutic.

In this book, Dr. Theuer and his colleagues address the dilemma of the rising cost of clinical trials and propose an alternative model to reduce these costs. The book is written by industry veterans who have worked in large pharmaceutical and small biotechnology companies and have first-hand knowledge and experience in advancing therapeutics through clinical trials to approval. The authors' experience in both settings informed them on the enormous cost of developing novel drugs. This experience helped them to propose a new model for clinical trial execution that reduces the cost of drug development. After reading the book, I am convinced the time has come to implement this approach. I am delighted with this book's goals, as I am sure it will be very useful for anyone in the drug discovery and development field. As more and more new therapeutics are discovered both in academia and industry, companies who adopt the business model outlined in *Unnecessary Expense* will play increasingly prominent roles in developing new drugs while also containing costs.

—Moorthy Palanki, PhD
Program Director
Professor of Practice in Biotechnology
College of Science, Technology, Engineering and Mathematics
California State University San Marcos
333 Twin Oaks Valley Road
San Marcos, CA 92096-0001
https://faculty.csusm.edu/mpalanki/index.html

INTRODUCTION

The commonly accepted cost of $1 billion to discover, develop, and approve each new drug is no longer accurate. That is startling, given that the book *The Billion-Dollar Molecule: The Quest for the Perfect Drug* was published less than a decade ago. Today, research and development costs have increased to more than $2 billion to secure FDA approval for each new drug. Most biotechnology and pharmaceutical companies are unable to support that level of investment on their own. Because the financial investment required is so extreme, drug-development companies understandably focus on discovering or developing drugs designed for large patient populations with diseases experienced by millions, such as high blood pressure, diabetes, or some of the more common cancers.

Unfortunately, this hamster wheel of chasing profits to justify investment costs stifles drug innovation. It's easy to see why. There's far less incentive to develop drugs for smaller patient populations; while the cost to develop and gain approval of a new drug for a rare disease may be the same as it would be for any new drug (approximately $2 billion), the profits are likely far less.

However, there is a new paradigm shift underway: an innovative drug-development strategy that decreases drug-development costs and shortens the time needed to test new drug candidates, while maintaining the rigor necessary to secure FDA approval. Domestic and foreign-based biotechnology and pharmaceutical companies who partner with TRACON to access our unique Product Development Platform (PDP) do so precisely to reduce costs and timelines without sacrificing quality. The results of greater efficiency allow for the development of drugs for underserved populations with limited treatment options. Moreover, faster and more economical drug development allows for a greater return of value to key stakeholders: patients in need who benefit from potentially life-transforming drugs, drug developers that receive equitable profits that reward their ongoing commitment to innovation, and insurance companies that prefer not to pay premium drug prices for drugs approved for patients with rare diseases.

In this book, you will learn what drives the high costs of developing new drugs; more importantly, you'll learn that many expenses are *not* the necessary cost of doing business. In fact, when assessing the current paradigms of drug discovery, development, and commercialization, it will become clear how the pharmaceutical ecosystem can benefit from adopting far more streamlined and cost-effective processes that empower and reward pharmaceutical companies that invest in the development of drugs for rare and common diseases, not just those that affect millions of patients. As you will learn in this book, the high costs associated with testing drug candidates in clinical trials encompass vast inefficiencies. You will learn about how investing in systems and processes obviates these inefficiencies and enables speed and quality when compared to the traditional path taken by many Big Pharma firms.

This book will also explain how dovetailing efficient drug development with the global harnessing of pharmaceutical innovation can greatly benefit the US population to the point that a drug candidate may be developed and approved for less than $20 million, which is one-hundredth the typical investment needed for the approval of each new drug. A significant paradigm shift offers a path forward for more companies to adopt better practices, either directly or through partnerships, and to broaden the development of the spectrum of drug candidates for the benefit of patients and innovators.

The goal of this book is to throw into sharp relief why more pharmaceutical and biotechnology firms should embrace a model that leverages global innovation and streamlines clinical trial execution for patient populations with high unmet needs to provide a faster path to profitability. By shifting the drug-development paradigm, TRACON has expanded commercial targets to include rare diseases with every expectation that we will generate a more-than-satisfactory return on investment. The TRACON PDP also serves as a solution for foreign-based drug-development companies looking to optimize the commercial potential of their drug candidates in the US market, which produces the additional benefit of utilizing global innovation for US patients rapidly and efficiently.

> BY SHIFTING THE DRUG-DEVELOPMENT PARADIGM, TRACON HAS EXPANDED COMMERCIAL TARGETS TO INCLUDE RARE DISEASES WITH EVERY EXPECTATION THAT WE WILL GENERATE A MORE-THAN-SATISFACTORY RETURN ON INVESTMENT.

ORGANIZATION OF THE BOOK

Like building a house from the foundation upward, each chapter in the book builds on the previous one, and certain chapters play to a common theme. Chapter 1 defines the billion-dollar drug problem and the factors that contribute to the high expense required to approve a new drug. Beginning with chapter 2, I chronicle my experiences as a cancer surgeon and then as a director of clinical development at biotechnology firms as well as in Big Pharma. In chapters 3 and 4, you will understand, through my journey in drug development, my "lessons learned" and how those lessons informed the solutions in place at TRACON today. Chapter 3 details my journeys in biotechnology firms and profiles one of the most successful cancer drugs, Rituxan®, which was developed at the first biotechnology company I joined, as well as less successful drugs, to reveal differences that account for successes or failures in drug development. This theme continues in chapter 4, where I review my experience at Big Pharma and the successful development of the drug Sutent®, which revolutionized the treatment of kidney cancer yet failed to demonstrate activity in other cancer types.

Chapter 5 exposes one of the prime business practices that contributes to the high cost of drug development: outsourcing the conduct of clinical trials to contract research organizations (CROs). You'll understand that CROs are not optimally aligned with pharmaceutical companies seeking quick, high-quality, and low-cost clinical trial execution, since CROs operate on a fee-for-service plus guaranteed payment model, meaning they are paid for every service they perform, whether or not that service actually contributes to the overall quality and execution of the clinical trial, and are paid a monthly fee, regardless of performance or work quality. Before moving to chapters 7 and 8, which present the TRACON PDP of CRO-independent clinical research as an antidote to the billion-dollar drug problem, you'll review in chapter 6 the critical partnership between biopharmaceutical companies and the FDA needed to navigate the complex regulatory process required for drug approval.

Chapters 7 and 8 examine the TRACON PDP as a means of decreasing the costs and timelines of drug development, while enhancing the quality of trials and data. You'll further learn that the alignment of interests between pharmaceutical partners through profit-share agreements that provide access to the TRACON PDP presents a superior solution for companies, especially foreign-based companies, to solve their problem of how to (1) access the lucrative US commercial pharmaceutical market, and (2) retain significant value of the drug candidates they have invested large amounts of time and money to discover and initially develop. The benefits of the TRACON PDP for foreign-based biotechnology companies serve America as well, by allowing us to leverage global innovation for US patients. Chapter 9 reviews the value of technology to enhance the efficiency and quality of drug development, and details how the TRACON PDP optimizes systems from technology providers for the benefit of the entire biopharmaceutical ecosystem.

Drug approval isn't the final step in the development of a new drug. Chapter 10 examines the finer points of commercializing an oncology drug, which differs markedly from marketing a drug prescribed by general practitioners. In closing we summarize the key takeaways and challenge all drug-development stakeholders to reevaluate current paradigms for the benefit of all members of the pharmaceutical ecosystem dedicated to advancing patient care.

Enjoy!

CHAPTER ONE

THE BILLION-DOLLAR DRUG PROBLEM

Until recently, it was commonly accepted that $1 billion must be invested to approve each new drug in the United States. While this may have been true in 2010, it is not true today. Capitalized research and development costs have been steadily climbing over the last ten years, to such an extent that now more than $2 billion in capitalized costs (including more than $1.4 billion in out-of-pocket costs and more than $500 million in fixed asset costs amortized over time) is invested to approve each new drug in the United States.

The approval of a new drug represents the culmination of a complicated and lengthy process that starts in a laboratory. There are six general steps that need to be accomplished to market a new drug (see Figure 1). Each step constitutes a successive cost increase, resulting in an overall developmental process that is very expensive and time consuming, with a typical drug approval requiring more than ten years from the time of initial dosing to patients in a clinical trial.

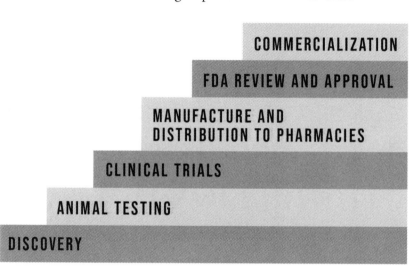

Figure 1: *General categories needed for the successful commercialization of a drug*

Most of the expenses associated with new drug approvals are dedicated to clinical trials (Figure 2). This exorbitance reflects both the high costs of conducting clinical trials and the high risk of new drugs failing to satisfy the stringent FDA safety and activity standards required for drug approval. The high risk of failure of new drug candidates is denoted by a common heuristic for drug discovery and development, the "one-in-ten" rule—that only one of ten drugs studied in clinical trials is likely to be approved.

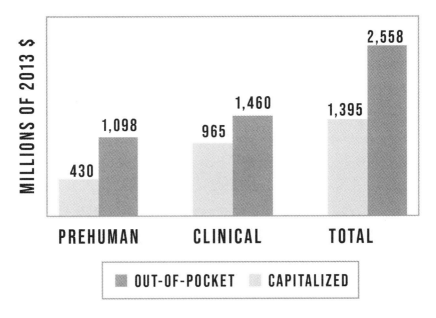

Figure 2: Prehuman phase, clinical phase, and total out-of-pocket and capitalized costs per approved new drug (Source: DiMasi et al., Journal of Health Economics 2016; 47: 20–33)

Drug development typically requires three phases of clinical trial testing in patients: Phase 1 trials primarily assess the safety of a drug candidate; Phase 2 trials primarily assess early signs of activity; and Phase 3 trials generally determine activity in comparison to an approved drug to see if the drug candidate is a better treatment. The "better" aspect differs based on the disease being treated. In the case of drugs to treat cancer, approval may be based on extending life span or reducing the size of a given cancer (also known as *generating a response*, something we'll go over in more detail in chapter 6 ("Understanding and Navigating the FDA")). Phase 4 trials look at drugs that have already been approved by the FDA. These studies typically assess safety over time in thousands of patients and may also explore other aspects of the treatment, such as quality of life or cost effectiveness.

As noted in Figure 3, approximately 60 percent of drug candidates are determined to be safe in Phase 1 trials and proceed to Phase 2 testing. However, Phase 2 trials, which are sometimes called the "valley of death," are the ones that are least likely to be successful due to a drug candidate failing to reproduce the activity demonstrated in animal models. Given a successful Phase 2 trial, a Phase 3 trial is also likely to be successful to allow for the filing of a New Drug Application (NDA) for a small-molecule drug, or a Biologic License Application (BLA) for a protein-based therapeutic, such as an antibody, with the FDA. However, there have been many high-profile Phase 3 trial failures, and even if Phase 3 trials are successful, not every NDA or BLA submitted is approved by the FDA, which adds additional risk (as well as time and cost) to each new drug approval prospect. If you multiply the conditional success rates of each phase of clinical development and the risk of FDA approval, the overall chance of approval for each drug candidate that enters clinical testing in patients is approximately one in ten.

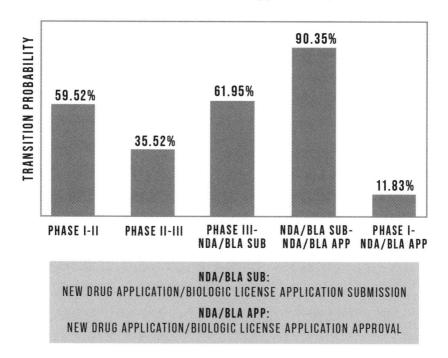

Figure 3: Probability of success for each step of the clinical trial process and overall, from clinical development success rates (Source: DiMasi et al., Journal of Health Economics 2016; 47: 20–33)

The risk of failure is even higher for cancer drugs, in which the overall chance of approval is closer to one in twenty, as noted in Figure 4.

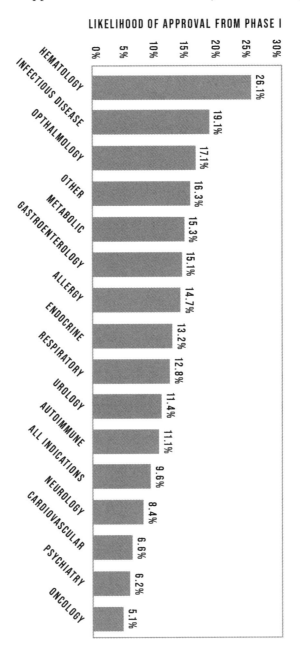

Figure 4: Overall probability of new drug candidate approval from Phase I by therapeutic area, 2006–2015, published by BIO

Figures 3 and 4 detail the risks associated with testing a new drug candidate in clinical trials as well as the risk of the FDA rejecting the conclusions that a potential new drug is safe and effective. Since an individual drug candidate typically costs hundreds of millions of dollars to study in clinical trials, and nine out of ten drug candidates are eventually discarded, the result is an aggregated cost of over $2 billion for each newly *approved* drug. Not shown are the even higher risks of failure associated with drug discovery and animal testing. If the risks of drug discovery and animal testing are included, far fewer than one in ten new drug "ideas" successfully navigate the arduous path of drug discovery, animal testing, clinical trials, manufacturing, FDA approval, and commercialization.

How can this business model even be sustainable? The answer is that an effective drug, commercialized successfully, can be extraordinarily beneficial for patients and for the company that brings it to market. Many successful cancer drugs have annual net sales in excess of $1 billion and are considered "blockbusters." These sales may persist for decades. One example is the drug Rituxan, a drug discovered by IDEC Pharmaceuticals, the first biotechnology company I worked at. Rituxan revolutionized the treatment of lymphoma and was subsequently approved for the treatment of autoimmune diseases.

Rituxan was first approved in 1997. More than twenty years later, in 2019, the last year for which annual sales data are available, Rituxan remains a standard-of-care treatment for patients with lymphoma and rheumatoid arthritis with annual sales in excess of $6.5 billion. All told, Rituxan sales have thus far exceeded $100 billion since its approval, a nice return on investment, and an estimated one hundred thousand patients have been treated.

The darker side of the drug-development story is that its high cost forces many biotechnology companies to marshal their resources in pursuit of a single drug candidate, effectively tying their fate to the success of one drug. Unfortunately, the high odds of failure inherent in the drug-development process will, more often than not, portend failure for these "single drug" companies. Another effect of the high cost and high risk of drug development is the need for a return on investment, which demands high net sales, which strongly biases companies toward the development of drugs for common diseases. This makes the development of drug candidates in rare diseases problematic, even if a rare disease population may be one most likely to benefit from new drug development. And those companies that do focus on rare diseases will likely charge a premium price to justify their investment.

That $2 billion price tag per approved drug is the same whether you're one of the top Big Pharma companies in the world or a small, innovative biotechnology company. Often, drug development is done in a hybrid manner: a smaller biotechnology company will discover a novel molecule and perform early clinical trials, and then a Big Pharma company will either license the drug candidate or acquire the small biotechnology firm, complete Phase 2 and Phase 3 trials, compile the dossier needed for FDA approval, and commercialize the drug. These were, for example, the steps taken during the approval of the drug Sutent, which I helped develop when I was at Pfizer. In that instance, the biotechnology company Sugen discovered Sutent and was purchased by the Big Pharma company Pharmacia, based on encouraging Phase 2 clinical trial data. Pharmacia, in turn, was then

> **THAT $2 BILLION PRICE TAG PER APPROVED DRUG IS THE SAME WHETHER YOU'RE ONE OF THE TOP BIG PHARMA COMPANIES IN THE WORLD OR A SMALL, INNOVATIVE BIOTECHNOLOGY COMPANY.**

purchased by the largest Big Pharma company in the world at that time, Pfizer (as of 2020, the largest Big Pharma company in the world is Johnson & Johnson; Pfizer is the second largest). Pfizer then recruited me and others to complete late-stage clinical trials that resulted in the approval of Sutent, a drug that revolutionized the treatment of patients with kidney cancer and became a blockbuster—more on that in chapter 4 ("Big Pharma Lessons").

SHIFTING THE PHARMACEUTICAL PARADIGM

Can the $2 billion price tag per successful drug be reduced significantly, allowing for a far more efficient process—one that offers greater incentive for innovation and focus on rarer patient populations with unmet needs?

The answer—unequivocally—is yes.

Many legacy systems within the pharmaceutical ecosystem are accepted as the "cost of doing business." And as you've learned in the preceding few paragraphs, that price is high, in cost and lost opportunities for patients with diseases for which drug-development investments aren't deemed sufficiently profitable. In reality, legacy processes in the biopharmaceutical ecosystem unnecessarily extend the time, cost, and risk of testing and marketing new therapies. The pharmaceutical paradigm can and must shift to a more agile model, one that emphasizes speed and thereby cost efficiency, without sacrificing rigor. By improving the overall efficiency of drug development, we can better promote the development of first-in-class (i.e., a drug with a novel mechanism of action) or best-in-class (i.e., a drug with a known and proven mechanism of action that offers superior safety, activity, or convenience compared to drugs already on the market) therapies for unmet need patient populations, reducing the overall cost of the journey from drug concept to commercialized reality.

At TRACON, we believe that we can secure FDA approval of a best-in-class drug candidate, one with the potential to revolutionize the care of patients with a rare type of cancer, on a budget of less than $20 million. We are currently employing this paradigm and expect this amount will be *the entire TRACON research and development budget required for approval.* At a less than $20 million investment, if approved, our drug candidate may be one of the least expensive drug approvals ever in the United States, especially in the oncology field, where clinical trials are complex and costly.

One key component is reducing the cost and time of the clinical research needed to prove the safety and activity of a drug candidate, while at the same time improving the quality of that research. At TRACON, we achieve this via our unique PDP, through which we dispense with one of the largest cost-of-doing-business components of drug development: that of conducting a clinical trial through a contract research organization, or CRO. We achieve this by implementing clinical trials using our internal team and streamlined technology platform that allows us to conduct efficient trials, with auditable and detailed data collation and reporting, at a greatly reduced cost. CRO incentives, unfortunately, are not optimally aligned with goals of pharmaceutical companies seeking quick, high-quality, and low-cost clinical trial execution, since they operate on a fee-for-service plus guaranteed payment model that does not encourage streamlined clinical trials. Instead CRO reimbursement creates the wrong incentives, which may generate behaviors that delay drug approval. I'll devote the entirety of chapter 5 ("CROs—The Pitfalls of Business as Usual in Drug Development") to the many drawbacks of the poorly aligned fee-for-service plus guaranteed payment model of CRO-based drug development and how the TRACON PDP makes this cost of doing business unnecessary. In fact, by reducing the cost and time of drug development, it is possible to more than double the value of each drug candidate. A doubling of the value of each drug candidate facilitates the development of twice as many new drugs for the same investment.

A second key is developing a drug with a proven mechanism of action that also has a unique product attribute, making it a potentially best-in-class therapy for an unmet need patient population. A validated mechanism of action lowers the clinical risk of failure, and a unique product attribute increases the value of the drug. Big Pharma is well aware of the value of best-in-class drug development, a topic we will further explore in chapter 4 ("Big Pharma Lessons"). One major issue with the current pharmaceutical paradigm, however, is that Big Pharma companies frequently invest in developing a best-in-class drug in an indication already well served by the first-in-class drug with the same mechanism of action in order to capture a share of the profit in that indication.

Instead of developing a drug with a similar mechanism of action in a large indication already well served by a first-in-class treatment, we advocate addressing patient populations without good treatment options. Because our development costs are lower, TRACON can expand our commercial targets to include many rare diseases and still generate a significant return on investment.

> **BECAUSE OUR DEVELOPMENT COSTS ARE LOWER, TRACON CAN EXPAND OUR COMMERCIAL TARGETS TO INCLUDE MANY RARE DISEASES AND STILL GENERATE A SIGNIFICANT RETURN ON INVESTMENT.**

A third key is leveraging worldwide investment in pharmaceutical research and development to benefit American patients, who represent 80 percent of the global pharmaceutical market. If a company in China, for example, has performed clinical trials in China and managed the complex industrial processes needed to manufacture a drug safely and reproducibly, then those learnings and that investment can be leveraged to benefit patients in the United States with similar or related diseases. This approach eliminates the cost of de novo discovery, manufacturing, and animal testing since those essential elements of drug development have already been accomplished by a foreign-based company in their native country. Furthermore, the available clinical trial data often reduces the size and scope of trials in US patients required for FDA approval. The process can then come full circle, because when a drug is approved for use in the United States, it usually enjoys a rapid path to approval and usage worldwide.

While the majority of drug innovation occurs in the United States, innovation in Asia and Europe has accelerated in recent years. This is nicely exemplified by the drug Opdivo®, one of the first approved therapeutics of a new class of drugs called immune checkpoint inhibitors (ICIs), which have revolutionized cancer care by allowing the patient's immune system to attack his or her cancer. You'll read a lot about ICIs in this book, as they represent one of the most important classes of cancer treatment that are now approved to treat nearly twenty different cancer types. The technology behind Opdivo was licensed by Bristol Myers Squibb from the Japanese pharmaceutical company Ono, which had earlier licensed the enabling patents from Kyoto University in Japan, where seminal research on this drug class was conducted.

TRACON works to identify unmet need patient populations in the United States and then subsequently conducts a global search for drug candidates to fulfill those unmet needs. Our PDP allows us to devise an efficient regulatory path for approval, navigate and secure regulatory concurrence from the FDA, and rapidly execute high-quality clinical trials at low cost. Finally, we possess the expertise to commercialize a new drug in the United States. These capabilities can provide a rapid and inexpensive path to approval and global commercialization.

We engage with our partners through a synergistic model whereby all parties are aligned, by sharing the risks and costs of drug development as well as the profits expected following successful commercialization. We also make clear that a dossier acceptable for drug approval in the United States for an unmet need patient population will likely also be acceptable for approval in multiple countries outside the United States. This business model, built using our CRO-independent PDP, is shifting the paradigm for effective US drug development. Since 2016, we have formed partnerships with multiple pharmaceutical companies in the United States and China to conduct trials and facilitate FDA approvals with the goal of benefiting cancer patients without effective treatment options.

CHAPTER TWO

JOURNEY FROM ONCOLOGY SURGEON TO BIOTECHNOLOGY EXECUTIVE

Before my two-decade career as a biopharmaceutical executive, I trained and worked as an oncology surgeon for more than a decade, with a two-year foray as a basic science research fellow at the National Cancer Institute (NCI). Mine is not the typical background for a biotechnology executive. My background as a cancer researcher and surgeon, however, has provided me with experiences that inform the impact and success I've had since entering the biopharmaceutical field.

As is true for many in the drug-development world, my entrée into biotechnology started with basic science research. During my college summers, I worked as a research assistant at Biogen, as part of a team endeavoring to isolate and sequence Factor VIII, a key clotting factor that is deficient in hemophiliacs. The goal of our project was to clone the Factor VIII gene to allow for mass production of the protein for hemophiliacs to prevent the bleeding complications that are characteristic of the disease. At that time, Biogen was a small biotechnology company. Fast forward to 2020: Biogen is now one of the largest biopharmaceutical conglomerates in the world.

I attended MIT for college, where I studied molecular biology, with a particular focus on understanding cancer pathways. From the very beginning, my inspiration and efforts focused on innovative oncology treatments. A paper I wrote as part of a graduate course on cellular immunology focused on a new drug class called antibody-drug conjugates (ADCs), also known as immunotoxins, and allowed me to deepen my understanding of how antibodies could be weaponized to attack specific cells.

Antibodies are natural proteins that bind to specific target molecules on the outer surface of cells with high precision. This specificity can be leveraged to create "magic bullets" consisting of an antibody fused to a toxic molecule or payload that kills *only* the cells to which the antibody binds. If the antibody binds to a receptor expressed only on cancer cells and not normal cells, the ADC becomes a magic bullet that selectively kills malignant cells, while sparing healthy tissue. That knowledge came in handy years later, when I conducted research at the NCI on this specific type of revolutionary antibody-based cancer treatment.

LEVERAGING COMMUNITY IMPACT

I attended medical school at the University of California San Francisco (UCSF) during the height of the acquired immunodeficiency syndrome (AIDS) epidemic. At that time, AIDS claimed hundreds of thousands of lives annually in the United States—there was no vaccine (and to this day, there still isn't an effective vaccine), and there were no effective treatments to prevent human immunodeficiency virus (HIV)–infected patients from developing severe immunosuppression that resulted in AIDS. I volunteered at the clinical epidemiology program based at San Francisco General Hospital, where we studied how HIV was transmitted and how HIV-mediated immunosuppression enabled seemingly innocuous microbes to become deadly opportunistic pathogens.

Through my AIDS epidemiology work, I was able to contribute to public health measures that had a positive impact in San Francisco, nationwide, and globally. Our team published several articles identifying HIV infection as responsible for increased cases of tuberculosis that spread to infect organs outside the lung, and that these cases of tuberculosis in HIV-infected patients indicated immunosuppression qualifying as a diagnosis of AIDS.[2] These findings promoted widespread testing of tuberculosis patients for HIV infection, which in turn provided HIV-infected patients with faster access to clinical trials of new treatments as well as social service benefits.

Our research also identified needle sharing among intravenous drug users as one of the primary routes of HIV transmission. We

---

[2] R. E. Chaisson, G. F. Schecter, C. P. Theuer, et al., "Tuberculosis in patients with the acquired immunodeficiency syndrome," *American Review of Respiratory Disease* 136:570–574, 1987.

C. P. Theuer, "Tuberculosis in patients with human immunodeficiency virus infection: review of current concepts," *Western Journal of Medicine* 150:700–704, 1989.

C. P. Theuer, D. E. Elias, G. F. Schecter, et al., "Seroprevalence and clinical features of human immunodeficiency virus infection in tuberculosis patients in San Francisco," *Journal of Infectious Disease* 162:8–12, 1989.

learned the value of providing clean needles to addicts through a syringe exchange program to prevent the transmission of HIV that occurred through sharing used needles. This public health measure was adopted in San Francisco and other cities to curb the HIV infection rate. By 2008, more than seventy-seven countries worldwide had introduced syringe exchange programs to help curb the spread of HIV.[3]

During that time, I also witnessed the activism of the gay community in San Francisco, which created close relationships with the medical community and facilitated the testing of drug candidates for HIV. Community engagement and cooperation between patients, providers, and drug companies resulted in some of the fastest drug approvals for HIV-infected patients ever seen in any therapeutic area. That important lesson made a lasting impression on me and contributed to the paradigm model we've adopted at TRACON to bring new therapies to cancer patients in a minimal amount of time.

CAREER PATH

While in medical school, during my surgical rotation, I witnessed an operation that crystallized how a simple, definitive surgical intervention made a permanent, positive impact on a patient. A young man with a gunshot wound was rushed to the operating room at San Francisco General Hospital with a barely perceptible blood pressure. The surgical team immediately incised the abdominal skin and, while blood spilled on the floor, intently searched to find the source of the bleeding. The damaged and hemorrhaging vessel was quickly identified, isolated, and repaired. Two days later the patient was smiling in his hospital room and asking when he could go home.

---

[3] "Syringe Exchange Programs around the World: The Global Context," *GMHC*, 2009.

To me, that operation epitomized the power of surgery and decisive, thoughtful action. Medical students and surgeons are familiar with the phrases "A chance to cut is a chance to cure" and "Cold steel will heal." Those phrases speak to the ability of a surgeon to immediately influence the course of a patient's life for the better and drew me like a magnet toward the discipline of surgery, despite the fact my medical school experiences were geared toward infectious diseases through my HIV research. After obtaining my MD degree at UCSF, I completed my training in general surgery at the Harbor-UCLA Medical Center in Los Angeles, a classic county hospital, where the plethora of cases allowed me to quickly accumulate hands-on clinical experience. Harbor-UCLA has a strong academic program and boasts an incredible faculty of attending surgeons who mentored resident physicians while also granting the leeway required for our growth into confident, independent surgeons. Our chairman of surgery noted in a speech prior to my fifth and final year of clinical training, "As chief residents in general surgery, you are all now the most powerful people in this hospital." That statement created a profound sense of responsibility and understanding of the impact we would have on our patients. We were taught to treat each patient like they were a family member and do well by them. It's a lesson that I carry with me to this day.

> **WE WERE TAUGHT TO TREAT EACH PATIENT LIKE THEY WERE A FAMILY MEMBER AND DO WELL BY THEM. IT'S A LESSON THAT I CARRY WITH ME TO THIS DAY.**

The Harbor-UCLA program is famous for its clinical training and clinical research, but it did not focus as intensely on basic science cancer research. Given my background conducting investigations in the molecular biology field at MIT and Biogen, I took advantage of the fact that one of my surgical resident colleagues had pioneered the path for Harbor-UCLA residents to become research fellows at the National Cancer Institute (NCI). I followed this path and applied for a two-year research fellowship at the NCI funded through the US Public Health Service.

During the interview process at the National Institutes of Health, I met with world experts in HIV and cancer research. These meetings almost led me to continue down a pathway of HIV-focused research, but my experiences as a surgeon underpinned my final conclusion that cancer research was a more logical choice, given that surgeons constitute a fundamental component of the cancer treatment community. With that clarity, cancer research made more sense.

I interviewed at several cancer laboratories, including the one led by Dr. Ira Pastan at the NCI. The paper I had written at MIT on immunotoxins was one of the main talking points during my discussion with Dr. Pastan, and my knowledge of how antibodies could be mechanized to attack malignant cells was instrumental in securing me an offer from the NCI.

OFF TO THE NATIONAL CANCER INSTITUTE

Dr. Pastan's lab at the NCI was incomparable in its productivity and the opportunities it presented to its research fellows. It published more papers annually than any other lab in the world at that time, and a new fellow could start working on important research projects from day one.

Through my work at the NCI, I learned an incredible amount about the process of developing a novel drug candidate. That process included the stages of initial drug discovery, testing in animals for activity and safety, drug manufacturing, and clinical testing including first-in-human trials (see Figure 1). This invaluable experience provided me with a deep knowledge of the therapeutic antibody space and reinforced my belief in the power of antibodies to treat cancer.

While I was not able to stay at the NCI long enough to witness an immunotoxin be developed into an approved cancer therapy, the first-in-class drug mechanism Dr. Pastan pioneered later became a standard-of-care approach for cancer treatment. This was accomplished through the efforts of Dr. Pastan's lab and through the endeavors of one of his fellows, Dr. Clay Siegall, who founded Seattle Genetics, a biotechnology company that has since successfully commercialized multiple ADCs.

After my two-year fellowship at the NCI, I had the opportunity to refocus my career on basic and clinical research. Directly interacting with patients daily, however, provided me with a great sense of satisfaction, and I returned to Harbor-UCLA to complete my surgical residency. I then joined the faculty at the University of California, Irvine (UCI) as a cancer surgeon, or surgical oncologist, where I committed myself to fighting cancer, one patient and one surgery at a time.

THE WAR ON CANCER

Cancer is legion, and the treatments available to us to fight it are dwarfed by its enormity. Cancer remains the second most common cause of death in the United States (after heart disease) and accounts for more than six hundred thousand deaths annually. Unfortunately, everyone reading this book likely knows a friend or family member who has battled cancer. A protean disease, cancer manifests in outcomes that range from curable to unsurvivable. The variables that dictate a cancer patient's chances of survival include the type of cancer, the stage at which it is detected, the availability of effective treatments (such as surgery, radiation therapy, or drug therapy), and the patient's unique medical profile. Patient profiles are more than just folders of data. Each patient represents a skirmish in a daily battle that occurs at cancer centers all over the world, as part of an ongoing war.

> **PATIENT PROFILES ARE MORE THAN JUST FOLDERS OF DATA. EACH PATIENT REPRESENTS A SKIRMISH IN A DAILY BATTLE THAT OCCURS AT CANCER CENTERS ALL OVER THE WORLD, AS PART OF AN ONGOING WAR.**

Most people are familiar with medical oncologists, the front-line doctors in the war on cancer who treat patients with cancer medicines, such as chemotherapy. They are perhaps the most deeply dedicated, empathetic, and caring subset of all physicians. These doctors carry out their duties knowing many of their patients will not survive their disease. With modern medicines, however, even in cases of incurable cancer, medical oncologists can both prolong a patient's life and improve the quality of that life, providing a priceless gift of extra time for patients and their families.

Surgical oncology, my field, is focused on removing cancer, and therefore lends itself to more curative outcomes. I loved being an oncology surgeon operating on gastrointestinal cancers, such as stomach cancer, colon cancer, and pancreatic cancer. Imagine the feeling of meeting with a patient whose cancer cannot be cured with medication or radiation, offering them the hope of a cure through surgical resection, successfully operating to remove the tumor, and then being able to tell the patient their chances of being cured are so high that they should now be more worried about dying in a car accident than dying of cancer. Not many other professions provide that degree of immediate gratification on a weekly basis, and it was this life-changing aspect of cancer surgery that drew me back to the operating table time and time again.

Cancer resections are some of the more intricate and technically demanding operations, requiring meticulous attention to remove "only" the cancer, followed by painstaking efforts to reconnect parts of the intestinal tract, usually in multiple places. Each operation can take up to ten hours to perform and represents a team effort—a primary surgeon, an assistant surgeon, a scrub nurse, and an anesthesiologist who all work together as one, bound by the satisfaction and responsibility of positively impacting a person in need.

As a full-time faculty member of the Department of Surgery at UCI, surgical oncology was my primary focus. While I could cure cancer through surgery if the cancer was detected early enough to be suitable for complete excision, many cancers are detected past the point where surgery can be curative. In those cases, surgical intervention can still prolong life by itself or in conjunction with drug and radiation therapy, but outright cure is sadly beyond the reach of a surgeon's capabilities.

TRIPLE THREAT

Being an academic surgeon allowed me to make a broader impact on the oncology field than I made in the operating room helping one patient at a time. Academic medicine afforded me the opportunity to be a triple threat: a clinical surgeon, a teacher of medical students, and a researcher. It was the research in particular that presented me with the opportunity to make a wider positive impact on cancer care beyond an individual patient.

Paramount for surgical cure is the early detection of cancer before it has a chance to spread to other parts of the body. The most important prognostic factor for any cancer patient is the stage of disease at diagnosis, emphasizing the importance of early detection for optimal patient outcomes. It was this prognostic significance that resulted in early detection techniques becoming the focal point of my cancer research at UCI.

There are many available effective screening tests that detect certain cancers before they become incurable. For example, a colonoscopy is recommended for every American starting at age fifty, as it can detect colon cancer while it is at an early stage and therefore still curable. Even better, precancerous polyps can be removed during colonoscopy before the polyps become fully cancerous. I noticed in my surgical oncology practice that I was treating many young Black men with colon cancer, who developed the disease prior to age fifty. (Darryl Strawberry, the former New York Mets and Yankees outfielder, and Chadwick Boseman, the late actor, are examples of this unfortunate phenomenon—both developed colon cancer at age thirty-eight.) Clearly, standard screening practices of the time were insufficient to help this population, as screening those patients for colon cancer at age fifty would be far too late.

One resource that drew my attention was based on my epidemiology experiences at UCSF. UCI housed a database of all cancer cases from Orange, Imperial, and San Diego counties. The database was part of the Californian Cancer Consortium, which contained clinical data on each case of cancer in California over several decades. This was big data back in the 1990s, before big data analysis became a popular practice!

Treating young Black men with colon cancer inspired me to research the effects of race, gender, and ethnicity on cancer detection and mortality. Using the California Cancer Consortium database, our team determined that Black men are at highest risk of colon cancer, compared to women or men of any other racial or ethnic group, and therefore would benefit from initiating colon cancer screening at an earlier age. We wrote several papers advocating that Black men be screened for colon cancer before they turned fifty years old based on that determination.[4] Shortly thereafter, the American Gastrointestinal Association lowered the recommended screening age of Black men for colon cancer from fifty to forty-five, referencing our seminal article. This revised guideline has since allowed for the detection of colorectal cancer in more patients at a curable stage.

Through my research I was able to contribute to a positive change on medical practice, affecting a far larger group of people than I could ever treat individually. As with my prior research in HIV, the opportunity to make a global impact was alluring. The highly individual "one patient at a time" nature of surgical intervention means that surgery isn't a naturally scalable discipline. Also, as gratifying as surgery was, repetition in anything can lead to a feeling of ennui. While the epidemiology cancer prevention research I was conducting remained highly meaningful to me, I was increasingly saddened by patients who presented with cancer at such an advanced stage that no intervention—surgical, medical, or radiation-focused—could cure their disease.

---

4   C. P. Theuer, J. L. Wagner, T. H. Taylor, et al., "Racial and ethnic colorectal cancer patterns affect the cost-effectiveness of colorectal cancer screening in the United States," *Gastroenterology* 120:848–56, 2001, 24.

        C. P. Theuer, T. H. Taylor, W. R. Brewster, et al., "The topography of colorectal cancer varies by race/ethnicity and affects the utility of flexible sigmoidoscopy," *American Surgeon* 67:1157–61, 2001, 21.

        C. P. Theuer, T. Taylor, H. Anton-Culver, "Screening for colorectal cancer" (letter), *New England Journal of Medicine* 343:1652, 2000, 34.

        C. P. Theuer, T. H. Taylor, W. R. Brewster, et al., "Gender-specific racial and ethnic colorectal cancer patterns affect the cost-effectiveness of colorectal cancer screening in the United States," *Journal of the National Medical Association* 98(1):51–57, 2006.

PAST EXPERIENCE, FUTURE OPPORTUNITIES

Not many surgeons join biotechnology companies. In part, this stems from the fact that surgery tends to encourage a very narrowly focused perspective between a surgeon and patient. Surgeons can, and should, be narrowly focused on a specific patient and operation to ensure the best likelihood of success. Moreover, many surgeons are not possessed of a research background, which lends itself to integration into the research environment at the heart of most biotechnology companies.

However, a former medical school classmate working at the biotechnology firm IDEC knew about my experience with antibody development at the NCI and encouraged me to look at opportunities at IDEC. My background in antibody therapeutics at the NCI lab made me a natural fit at IDEC, which was a company at the forefront of the development of antibody-based technologies to treat cancer.

At the time, IDEC was focused on the development of Rituxan, an antibody drug that was revolutionizing the treatment of lymphoma. They needed physicians to oversee clinical trials of Rituxan and other antibodies that explored new mechanisms of action for cancer treatment. I fully understood the potential for antibody therapies to radically transform and improve care and outcomes for cancer patients, and I found the possibilities offered at IDEC particularly exciting.

I thought long and hard about what it would mean to change my career and transition into the biopharmaceutical industry versus remain a surgeon. By joining the drug-development industry, I would have the chance to test and deploy medicines that could benefit thousands of individuals with cancer who could not be cured through surgery alone.

Therein lies the draw and the satisfaction of pharmaceutical development. There are always trade-offs, however, and one is delayed gratification. Biopharmaceutical drug development is a slow process, and the journey from discovery to commercialization can span a decade or more. This represented a stark contrast to the curative operations I was performing on a weekly basis in the operating room. My decision to join IDEC was a difficult one, and I wrestled with the idea for a long time before reaching a conclusion. Ultimately, because of the potential new intellectual challenges and the chance to make a broader contribution to medical care in oncology, I decided to pivot my career to the biopharmaceutical arena.

CHAPTER THREE
BIOTECHNOLOGY LESSONS

At the time I became a biotechnology executive, the field was still relatively young, and I had serendipitously experienced some of it during my college summers as a research assistant at Biogen. However, IDEC was not the company that put biotechnology firmly on the map; that accolade lies with Genentech, founded by my MIT Sigma Chi fraternity brother Robert Swanson. Until the 1990s, almost all drug treatments were chemicals. Then Genentech demonstrated—for the first time—that complex proteins (including antibodies) could be mass produced and be highly effective drugs. Today, Genentech, now part of the Roche conglomerate, markets some of the most commercially successful antibody cancer treatments in the world—including the blockbuster drugs Herceptin®, Avastin®, and Rituxan.

A MEGABLOCKBUSTER THROUGH A POSITIVE PARTNERSHIP

Rituxan was discovered by IDEC Pharmaceuticals, which was led by another MIT graduate, William (Bill) Rastetter. I joined IDEC as a director of clinical development in 2002. Researchers at IDEC had discovered a novel antibody that targeted B-cells, which give rise to certain lymphomas. Their thesis postulated that depleting B-cells would effectively treat certain types of lymphoma without causing significant distressing or harmful side effects.

IDEC first had the insight, and then took the risk of conducting animal testing, manufacturing and assessing a first-in-class medication to treat lymphoma patients in clinical trials. Their efforts established the activity of Rituxan in patients with lymphoma in Phase 2 trials, which successfully navigated around the valley of death in the clinical trial sequence, as we discussed in chapter 1 ("The Billion-Dollar Drug Problem"). However, IDEC needed to partner with a larger company to gain access to the capital and commercial infrastructure required to complete clinical trials and successfully market Rituxan. IDEC therefore partnered with Genentech, which, by the early 1990s, had become a fully integrated biopharmaceutical company with an extensive sales force.

The IDEC-Genentech partnership was a great example of drug development executed in a hybrid manner, whereby a smaller biotechnology company discovers a novel first-in-class drug candidate, performs early- and late-stage clinical trials required for approval in the initial indication, manufactures the drug supply for the initial commercial launch, and writes and prosecutes the BLA that leads to the initial approval, and then the larger company completes trials in additional indications to expand the population of patients who can benefit from the drug, and scales the manufacturing process to fully support the US market and to expand globally. The drug was commercialized by both companies through a copromotion partnership.

How successful were the Rituxan trials and the IDEC-Genentech partnership? Rituxan was commercialized in 1997, first in the United States and then worldwide, as a lymphoma treatment and has been marketed for more than twenty years. Following its initial approval in lymphoma, Rituxan was subsequently approved for the treatment of leukemia and rheumatoid arthritis as well as other autoimmune diseases, a process referred to as "label expansion" (i.e., approval in additional indications; see Table 1). Rituxan became a blockbuster drug within five years of approval and, even more impressively, generated more than $6.5 billion in annual sales more than twenty years later.

Rituxan's label expansion success is a great example of the pharmaceutical industry adage that drug development only begins with the initial approval—the table below shows how Rituxan was approved for additional indications more than twenty years following its initial approval. This is because the risk of approving an already approved drug in a second indication is generally lower than the risk of initial drug approval, as the safety profile has been established and mechanism of action validated. Label expansion therefore presents a much higher return on investment for the pharmaceutical company. While the development pathway of Rituxan epitomized the value of continued development following initial drug approval, if the pursuit of additional indications isn't done thoughtfully, misguided trials dedicated to label expansion can be a significant waste of money, an example of which we will explore in detail in the next chapter, chapter 4 ("Big Pharma Lessons").

YEAR	INDICATION
1997	Non-Hodgkin's lymphoma (initial approval)
2006	Diffuse large cell lymphoma in combination with chemotherapy
2006	Rheumatoid arthritis
2010	Chronic lymphocytic leukemia
2011	First-line maintenance for follicular lymphoma
2011	Two rare forms of vasculitis
2018	Pemphigus vulgaris
2018	Waldenstrom macroglobulinemia in combination with Imbruvica
2019	Granulomatosis in children

Table 1: Rituxan approvals by year

CLINICAL DEVELOPMENT AND DISCONTINUOUS INNOVATION

IDEC hired me as a director of clinical development to oversee clinical trials. At that time, one emphasis of the company was the development and commercialization of Zevalin, another drug that had been recently approved for the treatment of non-Hodgkin's lymphoma, a disorder of antibody-producing B-cells that was the initial indication for which Rituxan was first approved. Zevalin had been conditionally approved based on a high response rate (i.e., the demonstration that Zevalin could shrink lymphoma tumors in a significant number of patients) through the FDA's accelerated approval process (which we will discuss in greater detail in chapter 6, "Understanding and Navigating the FDA"), and the FDA then required further clinical trial data to demonstrate a survival advantage. IDEC and its partners delivered those data, as Zevalin was shown to improve survival when administered to lymphoma patients after they responded to initial treatment with chemotherapy.

First, a few more words about the three phases of clinical trials specific to oncology drugs. Phase 1 clinical trials are performed primarily to assess the safety of a drug candidate. These include the first-in-human Phase 1 trial, which administers a drug candidate for the first time to a person. In the case of oncology trials, usually that person is someone with advanced cancer of any type, whose disease is resistant to all known effective treatments, and therefore has no other treatment options. While the main objectives of a Phase 1 trial are to assess the safety and distribution of the drug candidate in the body, and to determine the optimal dose for further trials, participation in the trial also offers a patient an opportunity to benefit from the potential activity of the drug candidate.

Phase 2 clinical trials treat patients with a particular cancer type using the optimal dose identified and determined in the Phase 1 trial. Safety is assessed, but the main objective of the Phase 2 trial is to demonstrate activity of the drug to the point of establishing proof of concept. In the case of cancer drugs, evidence of activity, such as a reduction in tumor size (i.e., tumor response), may be sufficient for FDA approval using the accelerated approval mechanism in a tumor type without effective treatment options. This is the process by which Sutent, a drug I helped to develop while I worked at Pfizer, secured approval, which we will discuss further in chapter 4 ("Big Pharma Lessons").

Phase 3 trials are the last step in the clinical trial sequence and constitute a trial that typically enrolls hundreds to thousands of cancer patients in order to unambiguously demonstrate the activity of a drug candidate. In cases where Phase 3 trial participants have no other treatment options, patients may be randomized to receive either the new drug or a placebo. In other cases, when there is a known and effective treatment, patients may be randomly assigned to receive the new drug or the current standard-of-care medication. Finally, the trial may be designed so that patients will receive an accepted standard-of-care medication with or without the new drug candidate (see Figure 5).

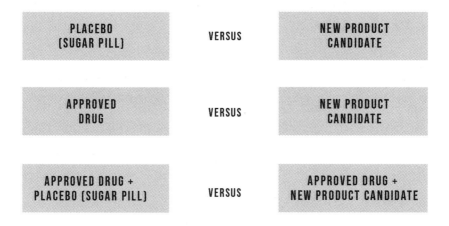

Figure 5: Typical Phase 3 clinical trial designs

Regardless of the exact design of a Phase 3 trial, enrolled patients are typically randomized into one of two arms so that patients in either arm will have an equal chance of developing side effects and responding to treatment. Furthermore, the patients and their physicians are usually unaware of which treatment is provided (known as "blinding") to mitigate any bias. Patients are monitored for side effects and outcomes, which might include the percentage of patients who respond to treatment with significant reduction in tumor size (i.e., have a response), the duration of time patients remain on treatment without evidence of tumor growth, or overall survival.

Zevalin is a complex and highly innovative therapy consisting of an antibody coupled to a radioactive particle. By specifically binding to cancer cells, the antibody component of Zevalin selectively delivers the radioactive particle to kill cancer cells while sparing normal tissues. Zevalin was the first approved radioimmunotherapy treatment and outperformed Rituxan in a head-to-head trial. Zevalin, however, did not combine as well with chemotherapy, an important standard-of-care treatment that Rituxan potentiated to promote patient benefit, resulting in a key Rituxan approval in 2006 in diffuse large cell lymphoma (see Table 1). Moreover, Zevalin treatment also required the engagement of nuclear medicine physicians, a group of physicians who are generally more comfortable participating in patient care by interpreting radiology reports than they are managing in the delivery of complex or novel cancer treatments.

These barriers resulted in poor utilization of Zevalin, despite its proven efficacy. Sometimes a drug is simply ahead of its time, similar to the way Ford Motor Company's Edsel, the "car of the future" of 1958, possessed a host of revolutionary features yet did not appeal to consumers. When approved in 2002, Zevalin represented a discontinuous innovation: its novelty did not reinforce current treatment paradigms or reinforce standard-of-care therapies. As a result, while Rituxan became a megablockbuster, with more than $8 billion in revenue in 2018, in comparison, Zevalin revenues in 2018 were a mere $7 million. It is important to recognize that Zevalin positively impacted lymphoma patients and continues to do so as a marketed cancer treatment nearly twenty years following its initial approval. But not all approved drugs prove to be commercially viable.

One example of the perils of commercialization is the drug Zurampic®, a drug that was developed to treat patients with gout. Typically, the body's kidneys dissolve and process any excess uric acid. Gout develops when uric acid accumulates and creates uric acid crystals that accumulate in the joint and cause pain. Gout can cause sudden and severe attacks of pain, as well as swelling and tenderness in the joints, most commonly the joint at the base of the big toe. Symptoms flare up intermittently and then remit. During a flare-up, the pain can be so severe that patients have difficulty walking.

Zurampic was initially developed by Ardea Biosciences, a San Diego biotechnology company, as a drug candidate to lower uric acid levels in patients with gout or at risk for gout. Following promising Phase 2 data, Ardea was purchased in 2012 by the Big Pharma company AstraZeneca for $1.2 billion. AstraZeneca then performed three separate Phase 3 randomized trials that may have cost up to $200 million each. One of these trials demonstrated that Zurampic increased the risk of kidney failure when given as a single agent and was terminated prematurely due to the increased occurrence of that side effect. However, the two other randomized Phase 3 trials showed that Zurampic, when combined with the approved generic drug allopurinol, lowered uric acid levels more than allopurinol alone. In 2015, Zurampic was approved by the FDA in combination with allopurinol for the treatment of elevated uric acid levels associated with gout in patients who did not respond optimally to treatment with allopurinol. In total, AstraZeneca may have spent nearly $2 billion to approve Zurampic (totaling the clinical development costs and the Ardea acquisition cost).

It therefore seemed surprising that AstraZeneca decided not to commercialize the drug, and subsequently sold Zurampic to Ironwood Pharmaceuticals in 2016 for $100 million and potential additional profits from success-based milestones totaling up to $165 million, in addition to a single-digit royalty on net sales. Why such a steep discount on the selling price? One possible explanation was the fact that Zurampic as a single agent was deemed unsafe due to the risk of kidney failure, which required a Black Box warning on the package label, and further concern that long-term treatment of Zurampic with allopurinol would uncover additional side effects. (A Black Box warning is the FDA's most stringent warning for drugs and medical devices on the market. Black Box warnings, or boxed warnings, alert the public and healthcare providers of serious side effects, such as injury or death.) Another possible issue was that the Zurampic Phase 3 trials proved the drug lowered uric acid levels in the blood but didn't prove the drug prevented gout attacks. Finally, another drug approved for the treatment of gout, Uloric®, was soon to come off-patent, thereby enabling the marketing of the generic version of Uloric at a much lower cost than that of Zurampic—in fact, the generic version of Uloric (generic name febuxostat) was approved in 2019.

While it may appear that Ironwood negotiated a great deal by acquiring Zurampic for a mere $100 million, in the end they, too, overpaid. Ironwood commercialized Zurampic in 2016, but sales were anemic (revenues struggled to top $1 million a quarter after more than a year on the market), and in 2019 Ironwood pulled Zurampic from the market in order to save $75 million to $100 million in annual expenses related to commercializing the drug.

The tale of Zurampic reinforces the point that approval of a drug doesn't guarantee commercial success or even a break-even return on investment. Likely due to safety concerns and unproven clinical benefits, as well as its cost compared to low-cost generic drug alternatives, Zurampic simply never established a market for itself.

Focusing on patient populations in need of new treatments due to the lack of effective treatment options is one way to avoid commercialization failure. It is also one way to avoid the waste of time, capital, and energy (i.e., opportunity costs) allocated to a drug that treats a disease already successfully treated by another drug, thus freeing up resources that can be allocated to the development and commercialization of drugs that truly matter by introducing a new type of treatment to patients who really need one.

THE BIG PICTURE

While my primary focus at IDEC was overseeing clinical trials of ADCs, I benefited most from the company's efforts commercializing Rituxan and Zevalin. IDEC, while a smaller company than Genentech, promoted Zevalin and copromoted Rituxan and was therefore a vertically integrated biotechnology company with all the infrastructure needed to bring a new drug to the market: discovery, manufacturing, clinical trial development and operations, and commercialization. Most importantly, these different departments were able to collaborate in an intimate way. Clinical development physicians could, and frequently did, walk down the hall to chat with the marketing team before meeting with the discovery team later that same day to discuss new drug targets.

Interacting with the commercial team was a completely new experience for me, and fortunately, members of the commercial team at IDEC were eager to share their extensive knowledge. In fact, the head of business development and marketing at IDEC at that time, Mark Wiggins, who also acquired marketing and business development expertise at Schering-Plough and Pfizer, is now the Chief Business Officer at TRACON. The ability to interact with Mark and his team at IDEC provided me with invaluable insight on how the whole pharmaceutical picture came together—from drug discovery to development to commercialization.

While Zevalin clearly wasn't becoming the commercial success we projected, it was my focus on trials of other drugs that I believed were unlikely to be approved that diminished my interest in continuing to work at IDEC. I left the company to work as a director of clinical development at Pfizer, the biggest Big Pharma company in the world at the time, on a drug that would transform kidney cancer care, an experience we will explore in detail in the next chapter (chapter 4, "Big Pharma Lessons"). Fortunately, this proved to be the right time to move on. Soon after I joined Pfizer, IDEC merged with Biogen, based largely on the value of the Rituxan franchise.

None of the cancer drug candidates in the IDEC pipeline were successfully developed.

My experience at IDEC was invaluable in that IDEC was small enough for me to appreciate the value of a vertically integrated company that successfully navigated each of the essential steps required for the commercialization of a new drug, in a setting where I was able to work intimately with key company stakeholders. I learned that drug activity (e.g., a drug binding its target on a cancer cell) doesn't guarantee clinical activity (e.g., prolonging life in a cancer patient), which doesn't guarantee drug approval by the FDA, which doesn't guarantee drug reimbursement and commercial success. Most importantly, I learned the value of co-development and co-commercialization relationships from Mark Wiggins, which he would later evolve into the aligned drug-development model that we propagate today at TRACON as a paragon of drug development and commercialization.

TARGEGEN

In 2004, the nimbleness and innovation of the biotechnology space drew me to a biotechnology start-up company called TargeGen. I joined TargeGen, a biotechnology company focused on the development of small-molecule drugs, as its Chief Medical Officer. At that time TargeGen was developing three early-stage first-in-class drugs for cancer and cardiovascular disease.

Early in my time at TargeGen, I read two seminal articles, one each authored by researchers in the United States and Europe, that delineated the cause of myelofibrosis, a rare form of leukemia, which induces anemia from fibrosis of the bone marrow, where red blood cells are made. The disease was a deadly one, and patients at the time were left critically underserved by the available treatment options.

These articles suggested many cases of myelofibrosis were caused by a mutation of the JAK2 gene, which resulted in the production of an aberrant protein that in turn activated a pathway, which caused the disease (see Figure 6 for JAK2 and BCR-*Abl* examples). The fact that a single gene mutation was responsible for cases of myelofibrosis was particularly important, as this contrasted with most cancers, where multiple gene mutations activate numerous cellular pathways that contribute to the disease. In the latter case, inhibiting a single pathway with one drug is often insufficient to meaningfully affect the cancer. Given that cases of myelofibrosis were driven by a single abnormal protein made from the JAK2 gene mutation, however, we knew a drug candidate that safely inhibited the function of the JAK2 protein would likely be very active. Therefore, the chance of a successful Phase 2 trial would be higher than for most cancer drugs.

*Figure 6. Examples of cancer types driven by a single gene mutation (BCR-*Abl *and* JAK2)

Big Pharma companies had not focused previously on targeting the JAK2 gene mutation that caused myelofibrosis. That meant there were no "off-the-shelf" drug candidates available for immediate testing in clinical trials when the seminal data were published indicating the JAK2 gene mutation to be the primary driver of most cases of the disease. I made the argument and convinced the TargeGen Board of Directors and senior management to refocus our team on the opportunity to discover and develop a drug to inhibit the JAK2 gene mutation.

Myelofibrosis is rare, with a prevalence of 16,000 to 18,500 patients in the United States. So why focus the company on creating a treatment for a relatively rare cancer? We took inspiration from Gleevec®, a drug marketed by Novartis, which is approved to treat chronic myelocytic leukemia (CML).

CML is a leukemia caused by the gene fusion protein BCR-*Abl*. As a result, a treatment targeting the pathway activated by that single gene mutation can effectively cure the disease. Kareem Abdul-Jabbar, the retired basketball star and writer, has been living with CML for years through successful treatment with the drug Gleevec, which targets the BCR-*Abl* gene mutation underlying CML. In doing so, Gleevec effectively "turns off" the disease and effectively turns a deadly disease into a chronic condition for nearly all sufferers. Gleevec's extremely high success rate (nearly every patient responds to treatment) translated into high sales, generating nearly $5 billion in annual revenue for Novartis prior to patent expiry that permitted generic competition, *despite the relatively small number of people with the disease*. The Gleevec story made clear that a highly effective drug, used by individual patients for years (and potentially for the rest of their lives, which in many cases equated to a normal life span), could be a blockbuster drug, even if the patient population affected by that disease was quite small.

The result of our discovery efforts on a JAK2 inhibitor to combat myelofibrosis was the drug candidate Inrebic®, which TargeGen advanced through Phase 1 and Phase 2 clinical trials. By that time, Inrebic was TargeGen's key value proposition, and in 2010 the company was purchased by the Big Pharma firm Sanofi for $75 million plus additional success-based milestones, which would total $560 million if the drug were approved and met commercial expectations. As TargeGen had functionally become a one-product company, the acquisition made sense, as it rewarded the TargeGen shareholders by allowing them an immediate return on their investment as well as future conditional payments. From the Big Pharma perspective, it was cleaner to simply purchase TargeGen rather than license the drug candidate Inrebic.

Sanofi completed a Phase 3 trial in 2013 that demonstrated Inrebic effectively treated myelofibrosis compared to a placebo. However, there was a problem: Inrebic caused irreversible brain damage in a small number of patients with vitamin B1 (thiamine) deficiency. While this side effect was rare, for some patients it was fatal. Also, by that time a first-in-class JAK2 inhibitor, Jakafi®, had been approved in myelofibrosis and successfully marketed by Incyte Pharmaceuticals. As a result, Sanofi decided not to commercialize Inrebic. This wasn't the end of the TargeGen story, however. Years later, my TargeGen colleagues were able to convince the FDA that Inrebic was valuable in patients with myelofibrosis despite the side-effect profile and sold the rights to Inrebic to Celgene for an even higher price than Sanofi paid to acquire TargeGen. The drug was then approved in 2019, more than five years after the Phase 3 trial demonstrated activity.[5]

---

5   Kyle Blankenship, "In a Boon for Buyer Bristol-Meyers, Celgene's JAK Med Inrebic Scores FDA Nod," FiercePharma.com, August 16, 2019, https://www.fiercepharma.com/pharma/once-left-for-dead-celgene-s-jak-inhibitor-inrebic-scores-fda-nod-eve-takeover.

The tale of Inrebic highlights a key issue (one of many) of Contract Research Organization (CRO)–dependent drug development—specifically, the quality of clinical trial oversight. (We will discuss this topic in greater detail in chapter 5, "CROs—The Pitfalls of Business as Usual in Drug Development.") The fatal brain damage seen with Inrebic, which arose through an exacerbation of a vitamin deficiency, was not clearly understood until the completion of the Phase 3 trial. It's possible a clinical trial program with better oversight and quality systems could have uncovered the issue earlier and remedied the problem by simply mandating that every patient take a multivitamin. (In chapter 9, "Integrating Technology as Part of the TRACON PDP," we will discuss how the technology part of the TRACON PDP, which is a core to our CRO-independent platform, could have helped mitigate this issue.)

The end result was a five-year delay in the approval of Inrebic and a worst-in-class safety profile that includes a Black Box warning for possible brain damage. In contrast, Jakafi has no Black Box warning. Given its superior safety profile and first-to-market advantage, Jakafi, which was approved in 2011, had sales of $1.7 billion in 2019 and is expected to reach $3 billion in sales annually in the near future.[6] In contrast, Inrebic was commercialized in 2019, and sales in the first quarter of 2020 were $12 million.[7]

---

6   Eric Sagonowsky, "Can Jakafi Hit $3B? Incyte Thinks So—and It's Got Other Launches on the Way," FiercePharma.com, February 13, 2020, https://www.fiercepharma.com/pharma/amid-prep-for-new-launches-incyte-aims-for-nearly-2b-from-jakafi-2020.

7   Keith Speights, "3 Reasons Bristol Myers Squibb's Revenue Skyrocketed in Q1," The Motley Fool, May 7, 2020, https://www.fool.com/investing/2020/05/07/3-reasons-bristol-myers-squibbs-revenue-skyrockete.aspx.

CHAPTER FOUR
BIG PHARMA LESSONS

After IDEC, I moved to Pfizer, the biggest pharmaceutical company in the world at that time in terms of annual revenue. Pfizer became famous for the mass production of penicillin during and after World War II and grew organically as well as through the acquisition of other companies. In 2003 Pfizer purchased another large pharmaceutical company, Pharmacia, in part for a drug candidate Pharmacia had acquired through an earlier acquisition of the biotechnology firm Sugen. Based on Phase 1 and Phase 2 trial data, that drug candidate appeared to have the potential to transform the treatment of kidney cancer and would become approved as the drug Sutent.

When Pfizer purchased Pharmacia, a significant amount of Pharmacia's clinical talent departed, possibly because they did not want to work at the biggest pharmaceutical company in the world. This created the opportunity for me to join Pfizer to oversee a clinical trial aiming to prove that Sutent could revolutionize the care of kidney cancer patients. That experience was obviously of great value, but of equal value was my exposure to best drug-development practices perfected at Pfizer, a company that set the gold standard on how to discover, develop, and commercialize new drugs.

At the time I joined Pfizer, the most commonly used drug for kidney cancer was Intron A®, which had poor activity, producing a lower than 5 percent response rate in patients, and was quite toxic, causing flulike symptoms with each administration. I joined Pfizer specifically to work on the registrational Sutent trial, which would deliver evidence of efficacy without undue adverse events and, most importantly, would be the basis for securing FDA approval.

Sutent's mechanism of action, decreasing the growth of cancer blood vessels and highly vascular cancers, indicated it should be most active in kidney cancer patients. The clinical data bore this out, as Sutent demonstrated a 30 percent response rate in a Phase 2 trial of kidney cancer patients, many of whom had already failed to respond to Intron A treatment. Evidence also indicated that Sutent was better tolerated than Intron A. That profile—a drug that is both more effective *and* safer than the current standard-of-care treatment—made it imperative to approve Sutent very quickly for kidney cancer patients.

In cases when a drug has the potential to radically transform the care of cancer patients, the FDA is an incredible ally (which we will discuss in greater detail in chapter 6, "Understanding and Navigating the FDA"). While many parties criticize the FDA, one of the best aspects of the current drug-development system in place in the United

States is that the drugs approved by the FDA are vetted more rigorously than anywhere else in the world. Once the FDA approves a medication to treat a specific disease, that drug's global reach is all but assured.

The FDA is eager to help bring transformational drugs to patient populations who lack highly effective treatment options. The oncology division of the FDA provides a path whereby initial approval can be based on response rate, rather than requiring a longer trial to prove a survival benefit. The idea is that response rates are seen rapidly, while proving a survival benefit may take years. In the case when a drug candidate is likely far superior to a currently approved therapy, neither the agency nor the drug company wants to deny patients that treatment during the time it may take to conduct a trial to prove a survival advantage. In those cases, the FDA allows the end point of response rate to serve as a surrogate end point for the expected survival advantage of the new drug in certain indications.

> ONE OF THE BEST ASPECTS OF THE CURRENT DRUG-DEVELOPMENT SYSTEM IN PLACE IN THE UNITED STATES IS THAT THE DRUGS APPROVED BY THE FDA ARE VETTED MORE RIGOROUSLY THAN ANYWHERE ELSE IN THE WORLD.

During the development of Sutent, I felt the FDA wanted to approve the drug as much as we did, given the current approved standard-of-care medication, Intron A, could have been considered a "toxic placebo," due to its low response rate and severe side-effect profile. The FDA agreed with us that demonstrating tolerability and a superior response rate for Sutent compared to the low historical response rate of Intron A would be sufficient for initial approval. The FDA allowed that a follow-up randomized Phase 3 trial could be done after the initial accelerated approval to prove a survival benefit by directly comparing Sutent to Intron A in a randomized Phase 3 trial.

Not surprisingly, Sutent demonstrated a greater than 30 percent response rate in the registrational Phase 2 trial and was quite tolerable. The drug was approved based on these data in 2006, and Sutent then doubled survival time compared to Intron A in the subsequent Phase 3 trial. Following approval, Sutent generated annual revenues of $1 billion annually for many years and was the most commonly used medicine to treat kidney cancer for more than a decade until even more effective drugs were developed called ICIs, which we will discuss in detail later in the book.

The Sutent approval process provided crucial lessons on how to reduce the cost of drug development. Sutent was approved far more rapidly than a typical cancer drug through the FDA's accelerated approval pathway. This was possible because Sutent targeted an unmet need indication of kidney cancer and demonstrated transformational activity. Clinical trial risk and time to approval are two of the most important factors driving the high investment needed to secure the approval of new drugs. Mitigating clinical risk by studying a disease where a drug is particularly active and decreasing timelines by focusing on an unmet need patient population that allows for accelerated approval are two powerful mechanisms that decrease the typical $2 billion investment needed to approve a new drug.

Beyond being part of a team that approved a drug that transformed the care of kidney cancer, working at Pfizer provided the opportunity to learn the best practices of drug development. Big Pharma companies may not be highly efficient, but they are unparalleled when it comes to the thorough investigation and testing of new drugs. In terms of drug development, Pfizer adheres firmly to the industry's highest regulatory standards. Moreover, Pfizer trains their employees extensively on best practices and educates every single person at their company to adhere to that standard. Many current biotechnology company executives have worked at Pfizer at some point in their career and have replicated those high standards at their subsequent places of work.

There is simply no better place than a Big Pharma firm to learn how to best conduct large-scale Phase 3 clinical trials, collate and submit data, and work with the FDA throughout the drug approval process. That paradigm also applies to the commercialization that follows drug discovery, development, and approval. Biotechnology firms may be far more nimble than Big Pharma entities, but if there is a valid criticism of biotechnology companies, it is that some firms do not check all the boxes. Pfizer checks *all* the boxes.

At TRACON we can confidently claim we check *all* the boxes throughout our drug-development processes as well. One reason we can make that statement authoritatively is because the majority of our executive team members have Pfizer on their résumés. Without that wealth of experience, we might not know exactly what boxes should be checked and the importance of doing so. Our deep familiarity with and commitment to the high standards practiced at Pfizer and demanded by the FDA are one of the many reasons domestic and foreign-based firms have chosen to partner with TRACON to assist in the drug development and approval processes.

Big Pharma firms have robust infrastructure, including access to experts in patient-related outcomes who assess the effect of drug treatment on a patient's quality of life, translational medicine to identify unique biomarkers on a tumor that might predict better drug activity, drug discovery, clinical trial execution, regulatory affairs, drug safety reporting, medical affairs, statistical analysis, commercialization, reimbursement, and more. In contrast, biotechnology firms may be able to execute drug discovery and oversee early-stage clinical trials, but in many cases, this represents the full extent of their capabilities. Even if clinical trials are successful, most biotechnology firms will sell themselves to Big Pharma or explore partnership options with Big Pharma firms for commercialization. This was the route pursued by IDEC and TargeGen, as you read about earlier in this book.

Commercialization expertise is critical to making life-changing or lifesaving drugs available to cancer patients worldwide. It isn't enough to simply publish a research paper and issue a press release announcing a newly approved and effective new drug. The marketing of medications is bound by almost as many rigorous rules as the testing of new drugs. Understanding your prescribing audience, patient preferences, and reimbursement practices is equally important to the success of a drug as its safety and efficacy profile, as I learned from the low adoption of Zevalin.

> COMMERCIALIZATION EXPERTISE IS CRITICAL TO MAKING LIFE-CHANGING OR LIFESAVING DRUGS AVAILABLE TO CANCER PATIENTS WORLDWIDE. IT ISN'T ENOUGH TO SIMPLY PUBLISH A RESEARCH PAPER AND ISSUE A PRESS RELEASE ANNOUNCING A NEWLY APPROVED AND EFFECTIVE NEW DRUG.

Big Pharma firms deploy thousands of marketing representatives to promote widely used treatments such as drugs that treat hypertension and diabetes. In contrast, most biotechnology firms simply cannot afford that kind of resource. A biotechnology firm can, however, commercialize a drug that is focused on a smaller patient population—such as those with rare cancers—and that is what TRACON plans to do (described in more detail in chapter 8, "Harnessing Global Innovation for US Patients"). Many biotechnology companies have successfully transitioned from discovery companies to development companies to commercialization companies (e.g., Genentech, Seattle Genetics, Exelixis, Celgene).

One downside of Big Pharma that isn't as common in biotechnology companies is siloed development. With so many people working on numerous different aspects of drug development and commercialization at Big Pharma, cross-departmental communication is often insufficient, so many employees are unaware of all that is going on, despite working on the same drug.

Another issue is that centralized processes may not prioritize key projects. At Big Pharma, drug-development project teams generally have to stand in line to gain access to internal resources, even if one drug is truly transformational. My goal at Pfizer was to advance Sutent to the front of the line for FDA consideration because all drugs are *not* created equal: Sutent was a medication that would have an enormous impact on patients, while also benefiting Pfizer's bottom line. While it may take an extra effort, it is possible to redirect the aircraft carrier that is Big Pharma. For example, Sutent made the cover of the Pfizer annual report in 2004, just over a year after the drug was acquired from Pharmacia.

One of the most frustrating centralized processes common to both Big Pharma companies (including Pfizer) and most biotechnology firms is outsourcing the conduct of clinical trials to contract research organizations, or CROs. The engagement of a clinical CRO is an accepted legacy practice and cost of doing business for most drug developers, domestic or foreign based. However, CRO engagements can, and often do, present significant impediments to efficient and innovative drug development. We will talk more about the pitfalls of clinical CROs in chapter 5 ("CROs—The Pitfalls of Business as Usual in Drug Development").

KNOW WHAT YOUR DRUG IS CAPABLE OF

The initial clinical trials overseen by the Pfizer clinical development team indicated Sutent, when given as a single agent, was a very effective treatment for kidney cancer as well as for a form of sarcoma called gastrointestinal stromal tumor (or GIST). As covered earlier when discussing Rituxan, drug development doesn't truly begin until the initial approval of a drug, and Pfizer looked to exploit this fact. The results of that effort provided another crucial lesson about inapt resource allocation in the pharmaceutical ecosystem.

Following the initial approval of Sutent based on response rate, we first proved Sutent to be the best drug treatment for newly diagnosed kidney cancer patients by comparing it head to head with Intron A. First-line positioning (treating a cancer patient as soon as they are diagnosed and prior to treatment with any other agent) is the optimal positioning for any cancer drug, as fewer and fewer cancer patients with widespread disease survive to receive subsequent lines of treatment. While Sutent was most effective in kidney cancer and addressed the high unmet need of those patients, it was also active as a single agent in patients with other tumor types who had progressed following treatment with standard-of-care drugs (usually chemotherapy).

This is generally the case: cancer drugs are typically more active in certain tumor types than others based on the mechanism of action of the drug. Many drugs initially developed for one type of cancer are subsequently approved for the treatment of patients with other tumor types whose diseases progress despite treatment with standard-of-care drugs for their diseases (i.e., as a "last-line" treatment). Establishing evidence of activity in the last-line setting of tumor types other than kidney cancer was one approach to expand the use of Sutent.

However, to reap even greater profits, Pfizer wanted to mimic the success of Avastin, a drug marketed by Genentech that inhibited the growth of cancer blood vessels via a mechanism of action that was similar to that of Sutent. One great advantage of antibodies like Avastin, however, is that they bind to one target and one target only, while targeted small molecules like Sutent are not as precisely specific, and therefore are associated with more side effects than antibodies.

Avastin was a very profitable drug by virtue of the fact that its side-effect profile allowed it to combine effectively with chemotherapy, a standard-of-care treatment for newly diagnosed patients with the common tumor types of colon cancer, breast cancer, and lung cancer. (Often cancer is treated with a combination of drugs so that if the patient's cancer continues to grow based on the development of resistance to a single drug, other drugs used in a combination may still work to limit cancer growth.) Avastin had been approved in combination with chemotherapy in newly diagnosed patients with all three of those common tumor types and therefore was quite profitable.

Pfizer wanted a piece of the profit pie Genentech had created with Avastin in those common cancers. And based on comparative data in kidney cancer, Sutent appeared to be the more active drug. Among other factors, Sutent was approved as a single agent in kidney cancer, while Avastin was approved in kidney cancer and other cancers only in combination with another agent. While single-agent activity may seem like an advantage on the surface, it was the ability of Avastin to combine with existing standard-of-care treatments in lung, breast, and colon cancer that propelled it to be a $5 billion oncology drug. Avastin, as an antibody, had a more benign side-effect profile than Sutent, a small-molecule drug. Given that chemotherapy is quite toxic, there was concern that the combination of Sutent with chemotherapy could be problematic.

Unfortunately, the temptation for first-line positioning in major cancer types proved to be too great, and Pfizer doubled down on Sutent to expand its application as a first-line drug by replicating the Phase 3 trial designs Genentech used to secure Avastin approval in breast, colon, and lung cancer. Sutent was combined with chemotherapy in each of these tumor types and compared to chemotherapy alone in large Phase 3 trials. Notably, all the trials failed. Sutent remained a very effective single-agent drug but was not approved in combination with chemotherapy. A strategy whereby Sutent was developed as a single agent in those large tumor types (e.g., breast, colon, and lung) may have better served patients and Pfizer.

Rather than respecting data indicating Sutent was an effective single-agent treatment without a high probability of combining successfully with chemotherapy, Pfizer spent what I would estimate to be $500 million to definitively prove the point. In the meantime, other Big Pharma companies approved their proprietary drugs with mechanisms of action similar to Sutent in patients following chemotherapy treatment. For example, Stivarga®, which is marketed by Bayer, was approved in 2012 for the last-line treatment of colon cancer patients. A key takeaway from that experience was the importance of fully understanding the drug profile and not asking more of a new drug than it can deliver. Another takeaway might be deciding when a new drug has been optimally developed and investing in others that can generate equal or better benefits.

THE PROFIT PIE

Another benefit of working at Big Pharma was learning the value of risk mitigation. As we've discussed, most new drugs fail. Internally developing, licensing or purchasing an unproven but promising drug requires a hefty investment, and Big Pharma firms invest in drugs likely to have the greatest profitability. Remember, at $2 billion invested to approve a new drug, those drugs that are successfully commercialized need to make several billion dollars in sales before patent expiration allows for generic competition and price erosion.

Big Pharma companies, therefore, hedge their portfolio of drug candidates by maintaining an active mix of different drug types in development at any given time. This mix consists of **first-in-class** (drugs with new and unproven mechanisms of action that represent the highest risk of failure but are also potentially the most lucrative); **best-in-class** (drugs with proven mechanisms of action that possess unique product attributes compared to the reference product—the ones I call "me-too-with-a-twist medications"); and **generics** (drugs that are identical to existing approved drugs with expired patents).

Big Pharma firms engage in organic research at their own discovery labs while also leveraging innovation that occurs outside company walls. They do the latter by putting their deep pockets to work, through licensing drug candidates or purchasing other drug companies, and then conducting the necessary large-scale trials to assess activity needed to secure FDA approval.

Big Pharma companies remain particularly alert to the financial benefits of joining a bandwagon based on the success of a drug from another company that validates a new mechanism of action. Incorporating the proven mechanism of a successful first-in-class drug into a new drug candidate dramatically decreases the clinical risk of failure. These me-too-with-a-twist drugs are medications with a primary mechanism of action that is nearly identical to the initial first-in-class drug but are chemically distinct enough to allow for patent protection without patent infringement. They also have the potential to improve the efficacy or safety profile of the first-in-class drug. This comparatively low-risk, high-profit endeavor ensures that multiple

Big Pharma participants have a seat at the table. In a sense, it's more cost effective for the pharmaceutical company to invest resources to circumvent another company's patent rather than manage the clinical risk associated with developing a first-in-class drug candidate.

What these "me-too" machinations mainly offer is a positive return on investment for Big Pharma companies, which are legally leveraging the investment and research and development work conducted by others. This practice rewards pharmaceutical companies without necessarily offering significant benefits to patients. That being said, me-too-with-a-twist drugs can move beyond simply carving out a piece of the profit pie established by the first-in-class therapy by gaining approval to treat patient populations not served by the initially approved first-in-class drug.

For example, in 2015 the FDA approved the Pfizer drug Ibrance®, a drug targeting the CDK4/6 pathway for the treatment of a specific population of women with breast cancer (those whose tumors were Her2 negative and hormone receptor positive). Soon thereafter, Eli Lilly and Novartis offered similar drugs that also targeted the CDK4/6 pathway. Each of these drugs—which initially were approved in the same breast cancer patient population—were "unique enough" in their chemical structures to prevent patent infringement. Indeed, the second- and third-to-market firms may claim that their version of a CDK4/6 pathway inhibitor have superior efficacy or safety profiles compared to Ibrance. Notably, however, the newer entries did not compare themselves to Ibrance for the purpose of gaining approval. Rather, they executed the similar trial design that Ibrance used to gain approval, by comparing their drug to a placebo.

In looking at the 2019 sales chart for competing breast cancer drugs (Table 2), you can determine for yourself the value of diverting resources to developing a me-too-with-a-twist medication. Pfizer's investment was clearly justified based on Ibrance annual revenues of more than $4 billion, which would be expected to persist for years. The sales of competing drugs are drastically lower than Ibrance, reflecting the first-to-market advantage of the first-in-class Pfizer drug. While the sales of the Eli Lilly and Novartis drugs may pale in comparison, what is noteworthy is that each is still a quite profitable franchise and will continue to be for years to come until the expiration of the patent allows for generic competition. Herein lies the rationale for the me-too-with-a-twist drug development path.

	PFIZER IBRANCE® ($ BILLIONS)	ELI LILLY VERZENIO® ($ MILLIONS)	NOVARTIS KISQALI® ($ MILLIONS)
Q1 2019	$1.13	$109.40	$91.00
Q2 2019	$1.26	$133.90	$58.00
Q3 2019	$1.28	$157.20	$123.00
Q4 2019	$1.28	$179.10	$155.00
FY 2019	**$4.95**	**$579.60**	**$427.00**

Table 2: 2019 revenues of approved CDK4/6 inhibitors

However, in some cases the twist can make a difference, which is true of Verzenio®, developed and marketed by Eli Lilly. Whereas

Ibrance proved its benefit in women with metastatic breast cancer, it did not prove to be active in women with early-stage breast cancer. In contrast, following its initial approval in women with metastatic breast cancer, the Eli Lilly product also demonstrated activity in women with early-stage breast cancer. Eli Lilly, to their credit, successfully addressed an unmet need patient population using a me-too-with-a-twist drug—a rare case in which a me-too-with-a-twist drug offers a distinct additional benefit to an unmet need patient population. Thus far the Novartis drug, Kisqali®, hasn't proven that its "twist" offers a benefit over Ibrance.

Until a me-too-with-a-twist drug demonstrates clinical benefit compared to the first-in-class drug, I refer to it as a "who-cares drug": a drug that is approved in the same indication as a previously approved drug that works by the same mechanism and has no demonstrated safety or efficacy advantage. Notably, this nomenclature reflects the patient's point of view: in the absence of clinical benefit, who-cares drugs aren't meaningful to patients, even though they may be highly profitable for the pharmaceutical company. However, to be fair, a me-too drug can have the effect of creating competition within a therapeutic area that lowers drug prices. For example, if there are multiple products with similar efficacy and safety profiles serving the same patient population, then pharmaceutical companies with the me-too product may try to undercut the price of the market leader in an effort to gain market share. That dynamic allows for a lowering of drug prices that can benefit patients, who frequently need to make copayments that supplement the reimbursement made to pharmaceutical companies by their insurance carriers. To date this dynamic has been more operant in the case of treatments for non-oncology indications, like hepatitis, rather than oncology.

Playing too much of the me-too-with-a-twist drug game can be viewed as an unfortunate waste of resources that may otherwise be dedicated to discovering first-in-class drugs or studying drugs with proven mechanisms of action to prove benefit in unmet need patient populations.

In oncology there are abundant unmet need patient populations with less common cancers in desperate need of more effective treatments. As an example, the most effective drug used to treat sarcoma patients is a chemotherapy discovered more than fifty years ago. Similarly, the most effective drug to treat brain cancer was also discovered more than thirty years ago. Regrettably, innovative treatments that address these rare unmet need patient populations may not be a priority for Big Pharma, as they may not present the blockbuster drug potential (i.e., a drug that generates at least $1 billion in annual net sales), which is the goal of most drug developers.

The financial rewards for me-too-with-a-twist drug development have created a self-perpetuating prophecy throughout the entire drug development ecosystem. If Big Pharma is more likely to invest in me-too-with-a-twist drugs, so will venture capitalists who invest in biotechnology companies with an eye toward selling the company to Big Pharma. As a result, these companies and other drug innovators are not necessarily incentivized to focus on the development of first-in-class medications. More and more resources are dedicated to copycat drugs focused on large patient populations, drugs that are just different enough or that can be aggressively marketed to grab a slice of that profit pie. Thus, the resources for the support of drugs with higher risk profiles (e.g., those with unproven mechanisms) may be limited. The result of this lack of support for innovation stifles the discovery and development of potential new first-in-class drug candidates that could successfully treat multiple indications.

Countless drugs die on the vine due to a lack of access to funds and resources required to support rigorous clinical trials. This is especially true at smaller companies with lower amounts of working capital. The innovation may be there, but the lack of financial resources inhibits the ability of smaller firms to study first-in-class medications adequately, especially within the United States, where the cost of CRO-managed clinical trials is often prohibitively expensive. Smaller, innovative companies almost exclusively focused on drug discovery likely lack the knowledge, relationships, and technology to execute clinical trials in the prescribed manner to fulfill the reporting requirements that allow a drug to effectively navigate the FDA and other region-specific approval processes.

> **COUNTLESS DRUGS DIE ON THE VINE DUE TO A LACK OF ACCESS TO FUNDS AND RESOURCES REQUIRED TO SUPPORT RIGOROUS CLINICAL TRIALS. THIS IS ESPECIALLY TRUE AT SMALLER COMPANIES WITH LOWER AMOUNTS OF WORKING CAPITAL.**

If drug development were conducted more efficiently, however, we could shift the pharmaceutical paradigm for the betterment of all stakeholders, ensuring that more drugs are developed profitably to benefit more patient populations. Consider this argument based on economic *and* ethical considerations: efficient drug development effectively aligns and rewards patients and pharmaceutical companies. Improving the efficiency of drug development would make the return on investment for a drug that addresses an uncommon cancer as high as the return on investment that currently exists for a drug approved to treat common cancers. For example, a $200 million investment in a drug that will potentially generate

$2 billion in annual revenues is a bet worth taking, but so is a $20 million investment in drugs that will potentially generate $200 million in annual revenues. In fact, supporting the development of multiple drugs at $20 million (or less) each better mitigates the risk of failure, while potentially increasing the benefits to thousands of patients that make up multiple unmet need patient populations.

In sum, inefficiencies in the drug-development process dictate a high profit threshold to warrant an investment by many Big Pharma firms. In turn, this prioritizes the development of drugs for larger indications. Efficiencies, however, can be improved by implementing several of the key themes discussed in this chapter:

- Know your drug to increase its chance of success (and don't try to force it to be something it isn't).

- Address patient populations with unmet needs to access the accelerated approval pathway, which decreases developmental timelines.

- Label-expand aggressively following initial approval as first- *or* last-line therapy.

There are additional proven strategies and tactics to streamline drug development that we'll explore in the next several chapters, including the following:

- Performing CRO-independent clinical trials to decrease the cost and time of development while also increasing quality

- Aligning development and commercialization goals between partners through profit sharing to maximize value for all stakeholders

- Harnessing global drug innovation

CHAPTER FIVE

CROS—THE PITFALLS OF BUSINESS AS USUAL IN DRUG DEVELOPMENT

Multiple tasks need to be flawlessly executed to successfully market a drug: discovery, animal testing, manufacture, clinical development (conducting clinical trials to demonstrate safety and efficacy), securing FDA approval, and commercialization. Each element in the process requires unique expertise and strict adherence to the highest possible standards of patient care, data management, and reporting to the FDA. At TRACON we specialize in executing clinical trials and mapping a pathway to approval to meet the high standards of the FDA. We also possess considerable commercialization expertise and engage with companies with innovative drug candidates who desire a fast-to-market strategy to successfully commercialize potentially transformational therapeutics.

We do this without employing clinical CROs to conduct clinical trials, which allows us to execute our trials faster, at a lower cost, and to higher quality than CROs typically retained by biopharmaceutical companies. To understand the advantages of the TRACON business model, we first need to understand the characteristics of clinical CROs that currently dominate pharmaceutical development.

The main function of a clinical CRO is to execute clinical trials, which can include early-stage (Phase 1), midstage (Phase 2), and/or late-stage (Phase 3) trials, in a manner that complies with the numerous FDA and other federal regulations. The state of the biopharmaceutical industry has evolved to the point where most Big Pharma and biotechnology companies outsource the conduct of all of their clinical trials to clinical CROs, with the justification that a CRO that exclusively focuses on clinical trial implementation in accordance with FDA regulatory standards represents an efficient use of resources. There are numerous examples that demonstrate precisely the opposite to be the case.

In theory, a CRO can sustain its business model by managing trials for multiple companies so that the CRO's resources can be switched to a second company's drug candidate in the event the first company's drug candidate fails, which we know happens more often than not. In contrast, a biopharmaceutical company with a limited pipeline that invests in the clinical operations infrastructure and personnel necessary to execute a clinical trial in-house may need to fire the entire clinical operations team in the event of a failed drug trial and will have wasted the up-front costs of establishing that infrastructure (e.g., codifying the policies and procedures of clinical trial execution, developing the data input forms and the database to store patient-related data from each trial participant, developing the safety reporting systems needed to ensure FDA compliance, and so forth).

Thus, the CRO model appears to be an attractive and logical choice for engagement by biopharmaceutical companies.

But.

CROs operate on a fee-for-service plus guaranteed payment model, meaning they are paid for every service they perform, whether or not that service actually contributes to the overall quality and execution of the clinical trial. In addition to this, they are paid a monthly fee *regardless* of performance or work quality. The result is that CROs are not optimally aligned with pharmaceutical companies seeking quick, high-quality, and low-cost clinical trial execution. Why would they be? They get paid regardless of performance.

An example from another industry familiar to many of us will help to illustrate the issues inherent in the clinical CRO reimbursement model. In a way, an auto repair shop can be considered a car repair CRO, minus the added benefit to the repair shop of a guaranteed monthly payment. For the purposes of this analogy, let's add in that CRO-like benefit and consider the following scenario: you take your car to an auto repair

> **CROS ARE NOT OPTIMALLY ALIGNED WITH PHARMACEUTICAL COMPANIES SEEKING QUICK, HIGH-QUALITY, AND LOW-COST CLINICAL TRIAL EXECUTION. WHY WOULD THEY BE? THEY GET PAID REGARDLESS OF PERFORMANCE.**

shop outside of any warranty insurance and meet the head mechanic, who provides a very straightforward estimate for the services required and assures you that he will take two days to complete the job. Three days later, you receive a call from a mechanic you've never met who tells you his boss couldn't do the job because he's been assigned to a more complex repair, and your job is taking longer to complete than originally promised. The new mechanic then tells you that while he repaired the initial problem, your car actually needs three or four additional essential repairs to be drivable, and by the way, those repairs will take an extra week to perform, and one more thing—there is a daily storage fee charged for storing your car at the shop until the repairs are completed, which started on the day you dropped off your car. The mechanics offer no warranty on the work performed but will gladly continue to let your car sit on their premises for the daily storage fee as you weigh your options of whether or not to move forward with the additional recommended repairs.

As you can see, the auto repair shop in our CRO comparison model has very little incentive to complete the job quickly and has every incentive to repair things that may or may not need to be fixed. They're happy for you to take your time to figure out what to do next, as you're paying for the time that your (currently undrivable) car sits on their lot. Given that it will also cost you money to move your car to another repair shop (that has the same reimbursement policy), you will also appreciate that you have very little leverage in this situation.

Now let's consider a hypothetical case of a clinical CRO. First, the CRO business development group will present to you, a potential biopharmaceutical client in the oncology space, their impressive capabilities, including résumés of monitors they employ. These monitors, the potential client is told, are the critical personnel who will ensure that all data from patients enrolled in your oncology clinical trial will be accurately entered into the database by the hospitals that treat the study participants with your drug candidate. The CRO representatives provide you with what seems to be a very reasonable estimate for the cost to conduct your trial and a short timeline to complete the trial.

You're convinced and sign on. As the trial proceeds, however, you first realize that none of the monitors whose résumés you viewed are assigned to your trial. Instead, the monitors assigned to your trial seem to know very little about the cancer field and have primarily worked on trials of drugs for urinary incontinence. Because their expertise is outside the field of oncology, they aren't aware of the complex evaluation criteria for your primary end point of tumor response rate, so they really don't have the knowledge base to determine whether a mistake was made by the site during data entry.

This is a prominent and common problem in the oncology trials space. Oncology trials are dramatically different from other drug trials. It's far harder for unskilled monitors to judge the effects of a new oncology drug based on response rate versus the effects of a new blood pressure drug, where it is very easy to interpret success: if a patient's blood pressure decreases to 120/70 from 140/90, the drug is working. Those are just numbers. But with oncology trials, analyses require far more discernment, and monitors with that type of sophisticated experience rapidly discover they can make far more money working independently and contracting directly with biotechnology or Big Pharma firms. Many monitors learn the basics of their business at CROs and then depart, leaving the CROs bereft of experienced personnel to manage and monitor more complex trials. This is especially true in the case of oncology clinical trials, where the demand for competent monitors is intense given the more than 13,000 ongoing cancer trials at the present time, as cataloged by clinicaltrials.gov, a website that lists ongoing clinical trials in the United States and elsewhere.

Not that that stops the CROs from touting their capabilities.

But wait—there are more surprises ahead.

You also notice actual costs are far higher than budgeted. Whereas the original CRO budget shown to you predicted that very few patients would develop serious side effects requiring a special report be filed with the FDA, in reality more than half of the late-stage cancer patients enrolled in your trial developed serious side effects that required hospitalization at some point, and you must pay a sizable fee for each incident report prepared by the CRO.

Moreover, you realize the trial is taking longer to enroll patients than expected. But even if no patients are enrolled, you're still paying a sizable monthly project management fee to the CRO, a fee that you started paying the second you signed the contract. You also notice that you owe a fee each time an inexperienced monitor issues a query to the hospital that treated a patient. Because the monitors lack experience, they are issuing many more queries than warranted compared to an experienced monitor who may be able to answer a question with a single well-worded query.

Further, despite using the exact hospitals recommended by the clinical CRO, and despite your repeated calls to the CRO, neither encouragement nor castigation seems to affect the pace of enrollment or the rising costs. While the laissez-faire attitude of the CRO renders them impervious to this carrot-or-stick game, one clear outcome is raising your anxiety level—and at least the CRO isn't charging you for that ancillary effect. When the trial is completed more than a year later than expected, at twice the anticipated cost, you breathe a sigh of relief until you discover that an audit of the hospitals that enrolled the patients indicates the data entered into the database are inaccurate and do not reflect what actually happened to the patients who received your drug candidate. What are you to do about unusable data?

The only way to ensure accuracy is to start over with a new CRO and new site monitors. At this point you are potentially trading one set of problems for a different set of problems.

As you can appreciate, the longer a trial takes, the more monthly project management fees a CRO collects, which disincentivizes the CRO to enroll patients quickly. The CRO fee-for-service model also emphasizes activity over accomplishment. As illustrated in the hypothetical example above, if a CRO bills for each query they issue to a hospital, they are incentivized to issue multiple queries, even when one or two well-worded queries would be enough to effectively resolve a question. If ten queries can be issued instead of one—the activity—then the CRO benefits financially. The overriding goal of answering the question quickly—the accomplishment—isn't achieved in a timely or efficient manner.

Given that CROs are paid regardless of results, in a sense, the CRO business model represents a misalignment of interests that reinforces incentives to overlook mistakes and correct them later through costly change orders, to search for services to amplify fees, and to delay trial conduct. This isn't to say that CROs are amoral, but the CRO business model is clearly not fully aligned and incentivized with the interests of their

> **THE LONGER A TRIAL TAKES, THE MORE MONTHLY PROJECT MANAGEMENT FEES A CRO COLLECTS, WHICH DISINCENTIVIZES THE CRO TO ENROLL PATIENTS QUICKLY. THE CRO FEE-FOR-SERVICE MODEL ALSO EMPHASIZES ACTIVITY OVER ACCOMPLISHMENT.**

biopharmaceutical company clients. Some may wonder why pharmaceutical companies tolerate such a unique payment model and don't choose to negotiate more favorable terms. The answer is that the competition to employ CROs is intense, given the more than ten thousand ongoing clinical trials of new oncology drug candidates alone. This makes it hard for pharmaceutical companies to introduce more favorable terms that align interests.

The ultimate nightmare, as illustrated in the example above, is when a company locked into a CRO agreement for an extended period receives a final study report of such poor quality that the results are unusable due to errors and omissions. This happens, and when it does, it can drive smaller, well-meaning biotechnology firms out of business.

Regardless of the sophistication of data analysis, *the primary data set must be accurate.* If not, it then becomes a garbage-in-garbage-out loop. Reputable pharmaceutical companies have suffered this fate and been forced to fire their original CRO, sometimes a smaller "boutique" organization that promised lower fees, and in turn hire a larger and more expensive CRO with a better track record to start over from square one. This represents a terrible waste of time and resources that could be better allocated to the development of innovative drug candidates. It's appalling on ethical grounds as well, as it means a promising drug is delayed from advancing to address an unmet need patient population.

Unfortunately, employing a large CRO doesn't guarantee quality. One issue with a small company employing a global CRO is the high demand for these CROs, as many are busy conducting trials for Big Pharma firms. This leaves smaller biotechnology companies with an undesirable choice regarding CRO selection. They can opt for a global CRO, recognizing that it entails high costs and that the team allocated to their project isn't likely to be the "A-team" of people assigned to work on Big Pharma trials. Or they can opt for a boutique CRO that bills itself as financially friendly but may only have the resources for "B-team" people or worse.

To mitigate this *known* risk of substandard performance by clinical CROs, most pharmaceutical companies hire their own clinical operations teams to police CROs that they in turn pay to actually conduct the clinical trial.

You read that correctly.

A company pays a CRO—an entity that is in the business of making a profit that is not fully aligned with the company's goals of efficiency, speed, and accuracy—and then *also* hires (and pays) an entirely new team within their company whose only job is to police the CRO. This is one reason why a typical budget for conducting an oncology trial of a drug candidate is more than $200,000 *per patient*. That number includes the CRO costs, the CRO profit margin, and the costs of the internal team to monitor the CRO to ensure data accuracy and efficiency.

WHEN CLINICAL CROS MAKE SENSE

Many large-scale clinical trials enroll thousands of patients to secure the data required for FDA approval. Few companies, including Big Pharma firms, have the resources to implement a global trial without contracting a clinical CRO. CROs can and will continue to play a critical role in drug development, especially if that drug candidate must be studied in global large-scale clinical trials. Sometimes, a large US-based CRO may even subcontract trials to Chinese and European CROs to access patient populations in these regions, with all the data integrated into the US-based CRO's database. Big Pharma companies can easily underwrite those costs, inclusive of those imposed on them by inefficient processes.

But early-stage and US-centric clinical trials are a different consideration. In the oncology field, Phase 1 trials typically involve three or fewer sites and are much more manageable than a global Phase 3 trial. Perhaps using a CRO makes sense for a "one product, one trial" firm that is confident of immediate acquisition following a Phase 1 trial. In that one-off scenario, it's pointless for the smaller firm to invest in an internal clinical trial team and systems required to actually implement the trial themselves—they may be best served by relying instead on an internal "shadow team" to closely police the CRO, in addition to employing independent contractors who act as site monitors to verify the accuracy of patient data collected at the hospitals administering treatment.

For biotechnology firms interested in staying in the game long term, however, it may be worth investing in the same data management tools that CROs use and, more importantly, investing in people with experience in using those tools. Investing in this way may facilitate the approval and commercialization of safe and effective drug candidates by creating an aligned model of drug development that also lowers overall time and cost of clinical trials.

THE BENEFITS OF BREAKING FREE OF CROS

In 2006, early in TRACON's history, our company managed many aspects of trials of our drug candidates in-house but outsourced certain tasks, including data management, to CROs. We always contracted independent and highly qualified site monitors that we knew from our collective years of working in the pharmaceutical field to verify that all data for each enrolled patient were accurately entered by the hospitals treating patients with our drug candidates. Many times, these professionals constituted members of A-teams at various CROs who recognized they could earn greater compensation by independently contracting with pharmaceutical companies.

After three expensive CRO-managed trials during our early years at TRACON, we realized being shackled to a CRO was inimical to our goal of successfully advancing potentially transformational drug candidates to market. The financial impact of working with CROs didn't lend itself to a sustainable business model, and we set out to redefine the model itself.

By 2011, we realized that by implementing systems that empowered our people to interact *directly* with clinical trial sites, rather than simply policing a CRO, we could conduct two trials for the price of a single CRO-managed trial. In a sense, we removed the middleman and, in doing so, eliminated vast inefficiencies and unnecessary expenditures. We call our CRO-independent clinical operations and development capabilities the TRACON Product Development Platform—PDP, for short. Through our PDP we've established a replicable clinical trial implementation process that combines speed with rigor in regard to testing, data collection and analysis, report creation, and the myriad other elements of clinical trial execution (see Table 3).

CLINICAL DEVELOPMENT	CLINICAL OPERATIONS	MEDICAL WRITING	DATA MANAGEMENT
Product development strategy	Site budget and contract negotiations (with legal) and site payment tracking	Investigator brochures	Electronic data collection and storage
Protocol design and development	Electronic trial master file document management	Protocols	Data cleaning
Medical monitoring	Timeline and budget development	Clinical study reports	CRF design and database set-up
KOL / Investigator engagement and recruitment	Reg doc collection / Site selection / Study start-up	Annual reports	EDC system training
	Plan and manual development	Investigational New Drug applications	Medical coding
	Central IRB submissions	Publications	Randomization
	ICF development		
	Site monitoring (contractors)		

BIOINFORMATICS	CLINICAL ANALYTICAL	STATISTICS	PHARMA COVIGILANCE
Hardware and software selection and management	Method development	Statistical analysis plan development	SUSAR report preparation
21 CFR Part 11 software validation	Sample management	Protocol design	Safety review committee meetings (internal and independent)
Software integration and development	Sample stability	DMC and independent safety committee participation	Narrative writing
Data analysis	Laboratory manual development	Data analysis	Safety management plan development
Programming and reporting (CSR, annual reports, interim analysis)			Signal detection
			Safety data exchange with partners

Table 3: Clinical operations managed by the TRACON Product Development Platform (PDP)

The TRACON PDP emphasizes quality through direct interaction between our team and hospital study managers to identify issues as early as possible. Unlike CROs, we are highly incentivized to resolve problems quickly and produce accurate databases that support our investigative and reporting efforts, as it's a far, far better use of our resources and time.

We don't make money if the trials we are funding ourselves take more time or are poorly conducted. We only make money based on sales revenue following FDA approval and commercialization of a drug. Delaying a trial simply delays our opportunity for eventual profitability.

There's an added benefit TRACON has brought to data collation and FDA submissions: we've utilized and developed technology to enhance efficiencies (something we will discuss in far more detail in chapter 9, "Integrating Technology as Part of the TRACON PDP"). One of this book's coauthors, Bonne Adams, TRACON's head of clinical operations, has spent years working with technology vendors, helping them understand what TRACON wants: a streamlined technology platform that allows for efficient data collection from hospitals in compliance with patient privacy guidelines and timely submissions to the FDA to ensure patient safety. Furthermore, the technological advances that technology vendors design to TRACON's exacting standards are available to other pharmaceutical companies who wish to contract with those same vendors. Our goal is the most effective interaction of data collection and collation platforms to avoid the duplication of manual effort and generate the highest-quality auditable reports for regulatory agencies.

The cost savings are dramatic. We conduct trials at less than half the cost of a typical CRO (approximately $100,000 per patient). And these cost savings don't include the cost savings gained by eliminating the need to redo trials that a CRO could potentially mismanage. We focus on conducting Phase 1 and Phase 2 proof of concept in clinical trials, which may be sufficient to secure FDA approval for unmet need patient populations, as we have discussed for Sutent in chapter 4 ("Big Pharma Lessons") and will discuss in regard to a new cancer drug candidate we are developing, envafolimab, in chapter 8 ("Harnessing Global Innovation for US Patients"). Moreover, we have also implemented Phase 3 trials in the United States and Europe using our PDP.

As you might expect, TRACON isn't the only company that has forsaken clinical CROs. Bonne and I were exposed to the CRO-independent model at IDEC. Seattle Genetics, which is captained by one of Ira Pastan's other protégés, Dr. Clay Siegall, also embraces a CRO-independent clinical operations model and touts their ability to execute clinical trials at a cost of $135,000 per patient, which is much lower than the typical CRO cost of more than $200,000 per patient. If trial quality and speed are not compromised, where then does this difference come from if not the CRO profit margin? It's worth considering.

To sum up: Instead of paying a CRO to conduct clinical trials and hiring and paying internal staff members to monitor a CRO, we have eliminated the middleman, hired in-house clinical trials personnel, and invested in technology systems to implement and oversee clinical trials directly. We designed and continue to propagate this CRO-independent model because there is a clear misalignment inherent in the fee-for-service and guaranteed payment system implemented by CROs. When you are paid irrespective of job quality or efficiency, there is no financial incentive to do a job quickly and well, and there is every incentive to permit a trial to drag on while collecting a monthly management fee.

By comparison, the TRACON PDP is wholly aligned with the needs of key stakeholders: drug innovators and patients. Our profit center is marketing an approved drug, not conducting clinical trials. Fast, efficient, and quality development are each essential components of the TRACON PDP so that our drug candidates reach market quickly and generate profits based on drug sales. We will further review the benefits of the TRACON PDP to our biopharmaceutical corporate partners in chapter 7 ("The TRACON Product Development Platform—An Aligned Drug-Development Solution"). First, let's discuss a biotechnology company's most important partner: the FDA.

CHAPTER SIX

UNDERSTANDING AND NAVIGATING THE FDA

Ultimately, the FDA is the single most important partner for any drug-development company. Given the importance of the FDA, I will expand on my views of that all-important entity. Throughout the book, I offer insight into some of the FDA's specialized paths to drug approval, the increased relevance of technology in ensuring best-in-class data collection and reporting data to the FDA, and TRACON's experience with the FDA in general and specific to envafolimab, a best-in-class cancer drug TRACON is developing in partnership with two China-based companies. This chapter outlines many of the ways one can secure FDA approval for a new drug or drug use.

The FDA is a highly structured and bureaucratic entity that serves a critical role in guaranteeing the safe provision of drugs to the public. Some wonder if the FDA is an essential government agency. The answer is an unequivocal yes and can be validated by understanding why the FDA was legislated in the first place. The FDA was created as a federal consumer protection agency with the passage of the 1906 Pure Food and Drugs Act. This law was the culmination of about one hundred bills over a quarter century submitted to rein in long-standing and serious abuses in the consumer product marketplace.

The 1906 act was passed thanks to the efforts of Harvey Washington Wiley, who at the time was chief chemist of the Bureau of Chemistry of the US Department of Agriculture, the FDA's predecessor, in response to the public outrage at the shockingly unhygienic conditions in the Chicago meatpacking houses that were described in Upton Sinclair's book *The Jungle*. Note the following emblematic passage from Sinclair's seminal work:

> *Jurgis talked with some who worked in the sausage-rooms, and who told him how now and then someone would lose a finger in the dangerous cutting-machines; and how when that happened they would stop the machine, but only for a minute or so; if they could not find the finger they would let it go and call it sausage.*

It was hoped that the new law would end food adulteration and quack remedies—the two major evils and targets of the twenty-year crusade near the turn of the twentieth century for the federal regulation of foods and drugs.

The FDA's mandate has expanded since then. For example, in 1938 that FDA was charged with regulating drug safety. Why? In 1937 the elixir sulfanilamide, manufactured and marketed by the S. E. Massengill Company, killed 107 people, including many children, because the company made the elixir using a poisonous antifreeze solvent, diethylene glycol. More than one hundred years later, the remit of the FDA includes oversight of all investigational drugs and approved drugs.

When interacting with the FDA, it is critical to employ an experienced and knowledgeable guide because FDA processes are complex, including the lengthy approval process. That experience and knowledge does not merely consist of knowing the correct forms and formats to use when submitting information, though that alone is an enormous hurdle. It requires seasoned judgment to enable the most productive interactions with the agency.

The FDA isn't going to change significantly anytime soon, nor should we want it to: it's an extraordinarily thoughtful, rigorous, and effective agency that strives to approve medications that are safe and can transform lives. Its stringent guidelines are tempered with reality and compassion, as evidenced by its willingness to accelerate the approval of drugs for unmet need or orphan patient populations where existing treatments make little or no difference. That's an especially important consideration in the oncology space.

The onus for gaining approval for a new drug (or new use for an approved drug) is firmly on the companies navigating the intricate and detailed FDA processes. Without expert knowledge, familiarity with format, and precise, pristine, auditable data, a company aiming for FDA approval will immediately encounter and experi-

> **WHEN INTERACTING WITH THE FDA, IT IS CRITICAL TO EMPLOY AN EXPERIENCED AND KNOWLEDGEABLE GUIDE BECAUSE FDA PROCESSES ARE COMPLEX, INCLUDING THE LENGTHY APPROVAL PROCESS.**

ence multiple opportunities to make mistakes, especially if they are participating in the approval process for the first time. Big Pharma and many biotechnology firms hire full-time regulatory affairs professionals to manage FDA interactions, while smaller firms may outsource these activities to a CRO. As you can imagine, at TRACON we feel direct interaction with the FDA works best, and the combined expertise of TRACON's management team with the FDA approval process is one reason why companies, both domestic and foreign based, are eager to partner with us.

ORPHAN POPULATIONS AND ACCELERATED APPROVAL

The FDA is committed to the rapid approval of therapeutics that advance patient care. The recent emergency-use authorizations of COVID vaccinations granted by the FDA are a testament to the dedication of the agency to ensure rapid approval of potentially life-saving therapies. In oncology and other diseases, the FDA offers a number of thoughtful, accelerated paths to approval, especially for new drugs designed to help unmet need patient populations. That doesn't mean the path to approval is any less rigorous in terms of the required clarity of data, reports, and easily auditable data trails. It means that the FDA will grant accelerated approval (sometimes called conditional approval) to commercialize a drug in the United States prior to completing, for example, a traditional randomized Phase 3 trial. Thereafter, upon the submission of more extensive trial data (i.e., randomized Phase 3 trial data) that demonstrate equal or greater efficacy of the drug compared to an approved drug already used to treat a given disease, the drug will be granted full marketing authorization. And from there, further approvals for expanded use tend to happen more rapidly.

The bar for accelerated approval in oncology can be a response rate of approximately 15 percent if those responses are durable, and the standard-of-care treatment has a very low response rate of less than 5 percent. While a 15 percent response rate may sound low, that represents a threefold improvement if the only existing treatment option has a 5 percent response rate. The FDA wants to see promising new treatments reach patients quickly. In these situations, response rate data from Phase 2 trials without evidence of a significant safety risk, rather than Phase 3 data, may allow for initial accelerated approval to grant patients access to the new drug. This process, for example, was very beneficial during the AIDS era, when the FDA worked closely with pharmaceutical manufacturers to bring new antiviral medications rapidly to market. It was also the process we utilized at Pfizer for the initial approval of Sutent in kidney cancer.

As we will discuss in chapter 8 ("Harnessing Global Innovation for US Patients"), the FDA's accelerated approval mechanism underpins our initial development strategy for envafolimab, whereby we expect that response rate will be the basis for initial approval of the drug. We have solid reasons to feel confident in our approach, as the FDA has an established history of approving new oncology medicines, including ICIs, based on response rate in unmet need patient populations with refractory cancers.

> **A NOTE ABOUT RESPONSE RATE:**
>
> Why is response rate the factor used by the FDA when considering early or accelerated approval? It's a consideration commonly employed for drugs in the oncology space. The efficacy of cancer treatments is best judged by their ability to prolong survival. However, trials to prove a survival advantage can take years to complete. Response rate (typically 30 percent tumor shrinkage, quantified on a CT scan) can be determined rapidly and, in many cancers, correlates with improved survival, as determined from prior clinical trials. Therefore, response rate serves as a surrogate end point for overall survival, the definitive and immutable end point. That's why the FDA is willing to consider response rate as the primary end point for consideration of accelerated approval.

In cases where Phase 2 trial data are sufficient for initial accelerated approval through a response rate end point, the sponsor will then generally need to conduct another trial (typically a randomized Phase 3 trial) to prove clinical benefit by showing a survival benefit. For example, after we approved Sutent at Pfizer based on response rate, we subsequently conducted a head-to-head Phase 3 trial versus the prior standard-of-care drug, Intron A, which demonstrated a twofold survival advantage. That was the basis for converting the accelerated approval into a full approval.

Securing accelerated approval is helped by input from key opinion leaders from the medical community. Often when there is a key FDA conversation, we need those unbiased opinion leaders on the call. Unbiased doesn't mean unopinionated; often these oncologists have been part of our studies in treating patients with new drug candidates. But they don't have a vested financial interest in a drug approval; their input is guided solely by their experience. If you're an oncologist treating a cancer patient who hasn't responded to any available treatment, both you and your patient are likely eager to participate in a clinical trial of a new drug candidate with the potential of an improved result. The feedback gleaned from early clinical trial participants—both oncologists and their patients—is both important and valuable to the FDA when the agency is considering accelerated approval of new oncology drugs based on a Phase 2 trial response rate.

Another way to extend the period of sales revenue is to secure orphan drug status. The Orphan Drug Act of 1983 is a law passed in the United States to facilitate development of drugs for rare diseases. The Orphan Drug Act was passed to encourage and provide special incentives for drug companies that undertake the development of drugs that target diseases affecting fewer than two hundred thousand people in the United States.

Under the Orphan Drug Act, drug companies can apply for Orphan Drug Designation, which, if granted, provides for exclusive marketing and development rights for seven years along with other benefits to recover the costs of researching and developing the drug. During that period of market exclusivity, the brand-name drug is protected from generic drug competition, regardless of whether the drug has patent protection. In addition to exclusive rights and cost benefits, the FDA will provide protocol assistance, a potentially decreased wait time for drug approval, and discounts or waivers on registration fees. Notably, the waiver of an NDA or BLA application fee is worth approximately $2 million. Another tangible benefit includes a tax credit of 50 percent of the qualified clinical drug testing costs upon drug approval. While there are significant benefits to securing an orphan drug status, this designation is not intended for drug companies to recover all the costs of drug development, but rather is intended as a cost reduction and regulatory streamlining mechanism to encourage and provide special assistance to companies that develop drugs for rare patient populations.

Despite the benefits, companies with orphan drugs in oncology and other therapeutic areas frequently feel that they need to charge a price premium to recoup their investment. While typical antibodies approved for cancer patients cost $10,000 to $12,000 for a monthly supply, orphan oncology drugs may cost three or four times this amount. For example, the orphan drug Folotyn®, marketed by Allos Therapeutics for the treatment of the rare indication of peripheral T cell lymphoma costs, more than $30,000 per month. Drugs marketed for rare tumors based on biomarker expression can command even higher price premiums. For example, the drug Vitrakvi®, marketed by Bayer Pharmaceuticals for the treatment of cancers with a specific fusion of the *NTRK* gene, costs nearly $400,000 annually to treat a single patient.

CHAPTER SEVEN

THE TRACON PRODUCT DEVELOPMENT PLATFORM

An Aligned Drug-Development Solution

TRACON has benefited greatly from our CRO-independent PDP to advance our pipeline of drug candidates through the clinical trial process. However, we also realized that our PDP, coupled with our commercialization expertise, could be of value to pharmaceutical companies lacking US commercial infrastructure that desired to commercialize their drug candidates in the United States. As noted in Table 4, a foreign-based pharmaceutical company with a promising innovative drug candidate has two unattractive options for US approval and commercialization: licensing their asset to a Big Pharma firm *or* contracting a clinical US-based CRO, maintaining full commercial rights, and then building a US sales force if the dug candidate is approved. We believe neither of these options is fully aligned with the best interests of the innovating company.

LICENSE TO BIG PHARMA	DO IT YOURSELF
• Big Pharma makes all decisions • Big Pharma may deprioritize commitment to program • Large, bureaucratic organizations that likely will outsource to a clinical CRO • Diluted long-term economics	• Requires a US-based team of experienced individuals to oversee the clinical US-based CRO • High cost with potential for cost overruns • High potential for clinical trial delays • Risk of a poor-quality clinical study report • Lack of control and no urgency or alignment with the CRO • Need to build a US sales force

Table 4: Typical options for a foreign-based pharmaceutical company to commercialize their drug candidate in the United States

Many times, drug-development companies located outside the United States that discover an innovative drug candidate will license it to a Big Pharma firm early in its development (e.g., prior to clinical testing or either during or following Phase 1 testing). By so doing, the company that discovered the drug is leveraging Big Pharma's clinical development expertise to design a series of clinical trials to provide evidence the drug is safe and effective, manufacturing expertise to ensure manufacturing is done to the high-quality standards mandated by the FDA, commercial expertise to market the drug, and regulatory expertise to ensure compliance with federal regulations both during the drug approval process and following commercialization. However, a price must be paid to access all these potential benefits and permit access to the US pharmaceutical market, which is the largest pharmaceutical market in the world. It's a high price, especially if the drug candidate is licensed to a Big Pharma company during an early stage of development.

Lack of access to capital, the desire to share the risk of drug development with a partner, and a lack of commercial expertise are three major factors influencing biotechnology firms both within and outside the United States to license an asset to a Big Pharma firm, typically for an up-front payment and success-based milestones, as well as a small portion of the profits as a royalty if the drug is marketed. However, the blended royalty paid by a Big Pharma firm for an early-stage asset will typically be approximately 10 percent, meaning that if the drug is successful, the innovative drug discovery firm will retain only a small fraction of the value of what could potentially be a blockbuster drug with annual billion-dollar revenues.

The Big Pharma firm will argue that the high risk of failure; high cost of clinical trials, manufacturing, and commercialization; and long timelines for drug development are factors that warrant that they retain the lion's share of the profits. As we now know, one reason for the long timelines and high costs is the typical reliance of Big Pharma on clinical CROs, and their misaligned fee-for-service plus guaranteed payment model. The same argument applies in the scenario where the Big Pharma firm acquires the asset by buying a one-product company.

In this case an up-front payment is typically made as well as contingency payments for FDA approval and certain sales revenue thresholds that are distributed to the biotechnology company shareholders if or when those milestones are achieved.

However, selling an innovative drug candidate to a Big Pharma firm doesn't guarantee that the drug candidate will advance into clinical trials. Once a drug asset is licensed or purchased by a Big Pharma firm, it is thrust into competition with all the other promising drugs within the Big Pharma firm's pipeline that were discovered internally or licensed from other pharmaceutical companies. While the asset may be the belle of the ball when licensed or purchased, it can rapidly become the ugly duckling if a higher priority asset is subsequently licensed. As well, the vagaries of large bureaucratic firms mean that the person determining drug priorities changes frequently. The next person in charge may decide not to prioritize clinical trials for the licensed or purchased asset. This all too often results in the drug candidate languishing on the shelf, and innovation is once again stifled.

Ambitious foreign-based firms with access to capital that are willing to fully accept the risks of drug development may hire a CRO to run trials in the United States (see Table 4). This presents a major logistical challenge. Policing a CRO from abroad without a US branch office to circumvent language barriers and time zone differences is problematic. Even foreign-based drug firms with US branch offices may not have the requisite experienced shadow teams to effectively police clinical CROs in the United States. The potential for problems is significant, and the quality issues, delayed timelines, and cost overruns that we reviewed in chapter 5 may cause the foreign-based innovative drug company to flounder. Even if these issues are overcome and the drug candidate is approved, the foreign-based firm still needs to solve the challenge of commercialization in the United States.

A BETTER SOLUTION

There is a better solution for a foreign-based pharmaceutical company eager to maximize the value of their innovative drug candidate than licensing to Big Pharma or managing a clinical CRO and commercializing themselves in the United States, one that allows them to share the risk, maintain a large portion of the value of their drug candidate, and work with a partner who is as motivated as they are to see their drug reach the market. As you might expect, through accessing the TRACON PDP, our partners can participate in a strongly aligned model of drug development to secure FDA approval and subsequent US and global commercialization.

TRACON's partners benefit from a far more efficient and cost-effective solution that affords a more equitable sharing of economics than a typical Big Pharma license. What we do differently at TRACON is employ a cost-, risk-, and profit-share model with our partners, customized for each, which creates strong alignment in our respective priorities while at the same time adhering to high standards at each step of the drug-development and commercialization process.

Our goal at a high level is to split the costs of developing and commercializing a drug candidate fifty-fifty with our corporate partner. Because our partner discovered the drug candidate and likely has committed substantial resources to manufacture that drug candidate, often TRACON will absorb most or all of the clinical trial costs (e.g., Phase 1 and Phase 2 trial costs, which may be sufficient for approval using

> **TRACON'S PARTNERS BENEFIT FROM A FAR MORE EFFICIENT AND COST-EFFECTIVE SOLUTION THAT AFFORDS A MORE EQUITABLE SHARING OF ECONOMICS THAN A TYPICAL BIG PHARMA LICENSE.**

the accelerated approval mechanism we reviewed in chapter 6) to assess the safety and efficacy of the drug. Since we execute clinical trials using the TRACON PDP, we spend our own capital to conduct the trials at a cost of less than half of what a CRO would typically charge to do the work. Also, because we fund and implement the trial ourselves, we have every incentive to conduct the trial quickly. Speed is important but not sufficient. Unlike a CRO, we expect to make a profit based on net sales and not based on fees for conducting the trial. We therefore need to conduct the clinical trials with the utmost rigor to ensure the data and the reports those data inform meet the stringent high standards of the FDA. In contrast to development utilizing a CRO, there is no detrimental misalignment of interest. Speed, low cost, and high quality are equal, critical, and aligned elements of the TRACON PDP.

As noted, in our partnership agreements TRACON generally manages and funds initial clinical trials designed for initial approval—in that manner we earn our way in; we then split the remaining development costs (e.g., for label expansion) and commercial costs evenly with our partner. We also split the profits, which grants our partners a far higher share of profits than they would likely receive from Big Pharma. For example, if the cost of commercialization is 20 to 30 percent of net sales and we split profits evenly with our partners, then they reap an effective 35 to 40 percent royalty. Compare that with the approximately 10 percent blended royalty rate typically paid by Big Pharma, and you can appreciate the difference in cash flow (see Table 5). We find that companies that believe in their drug candidates are particularly likely to appreciate the economic benefits of the TRACON partnership model, and as a result, will forgo an up-front fee in lieu of the significant potential back-end economic advantages presented in Table 5. If a potential partner believes their drug may fail, then a more traditional pharmaceutical deal that includes an up-front payment and success-based milestones may make more sense.

YEAR	SALES (IN MILLIONS)	REVENUES TO INNOVATIVE COMPANY THROUGH TYPICAL BIG PHARMA LICENSE (10% ROYALTY)	REVENUES TO INNOVATIVE COMPANY THROUGH TRACON PDP PROFIT SHARE (40% EFFECTIVE ROYALTY)
1	200	20	80
2	400	40	160
3	600	60	240
4	800	80	320
5	1000	100	400
6	1000	100	400
7	1000	100	400
8	1000	100	400
9	1000	100	400
10	1000	100	400
TOTAL	**8000**	**800**	**3200**

Table 5: Comparison of revenues in millions of US dollars over a ten-year period using a typical Big Pharma royalty versus the TRACON PDP for a blockbuster drug with a five-year ramp to peak sales

We modeled the risk-adjusted net present value (or NPV, a measure of the value of a product in today's dollars that takes into account the risks of each phase of drug development) of a drug candidate that could be immediately advanced into a registrational Phase 3 trial paid for by TRACON or paid for by Big Pharma paying a CRO. We assumed that in each case the drug would be approved and initially marketed in three years and would generate peak sales of $1 billion annually. Further, we modeled that achieving full market penetration to achieve peak revenues across multiple indications would take eight years. In the case of the Big Pharma deal structure, the partner is entitled to an up-front payment of $70 million, success-based milestones of $300 million, and a 10–14 percent blended royalty on sales. In comparison, in the case of the TRACON deal structure, there is no up-front payment or success-based milestone payments, and the partner is entitled to a profit split (or the equivalent of a 35-40 percent royalty).

Table 6 compares the risk-adjusted NPV to a partner of the typical Big Pharma deal structure to the TRACON profit-share deal structure. We present risk-adjusted NPV calculations based on various probabilities of success of the registration trial. The results indicate that at or above a 10 percent probability of success, the TRACON profit split deal structure generates superior value for the partner. Note that the success rate for a registration trial (e.g., Phase 3 trial) is typically in excess of 50 percent, as noted in chapter 1 ("The Billion-Dollar Drug Problem"), and few companies would initiate a registration trial with only a 10 percent probability of success. Thus, the TRACON deal structure has significant economic advantages compared to a typical Big Pharma license for an innovative company in need of a development and commercialization partner for their drug candidate in the United States.

POS	TRACON PDP	PHARMA LICENSE
10%	131	131
20%	262	176
30%	393	222
40%	524	267
50%	655	312
60%	786	357
70%	917	402

Table 6: Risk-adjusted net present value comparison of the TRACON deal structure and a typical Big Pharma deal structure based on probability of success (POS)

Our expertise in US commercialization is welcomed by foreign-based companies that aren't familiar with the marketing rules and conventions of the world's largest pharmaceutical market, as is our willingness to allow our partners to promote the drug with us (by employing their own sales force to sell the drug in collaboration with our sales force) to gain commercialization experience in the United States. Another advantage is that the dossier we file with the FDA for US approval can likely be used by our partners to file for approval both in their home territory and anywhere in the rest of the world, allowing them to rapidly market the drug in their host country and elsewhere. (TRACON focuses primarily on commercializing the drug in the United States, which represents approximately 80 percent of the global pharmaceutical market.)

As we've discussed, most drugs fail in clinical trials, with the greatest risk being Phase 2 trials designed to demonstrate initial efficacy of the drug. Because TRACON will fully fund clinical trial expenses through the Phase 2 proof-of-concept trial, our corporate partner benefits from the rapid determination of whether the drug candidate is likely to be approved entirely at our cost. Furthermore, if we split Phase 3 costs with our partner following initial approval based on Phase 2 data, they pay approximately one-quarter of what they would have paid a CRO to conduct the same trial, given our ability to conduct trials at half the cost of a CRO.

One final advantage of a TRACON partnership is the prospect of an integrated global trial. If our corporate partner can conduct clinical trials in their home territory, then we can collaborate to enroll patients from across the United States, Europe, and Asia. Accessing large numbers of patients will accelerate the conduct of clinical trials, and the TRACON PDP can be leveraged globally, anywhere competent clinical trial site monitors are available. While the FDA requires a clinical trial patient population that is representative of the intended US treatment population to grant marketing authorization, not every trial participant needs to be enrolled at sites in the United States, Europe, or Australia, where Caucasian patients represent the largest patient demographic.

THE ECONOMIC EQUATION: TIME, MONEY, QUALITY— WE CHOOSE ALL THREE

We've reviewed the value of the TRACON profit-share deal structure to increase the value of a drug candidate to our partners. But that is only half the value proposal we offer. How valuable can the TRACON PDP be in terms of enhancing the value of a drug candidate compared to CRO-based development? Let's review the three key categories of time, money, and quality. It typically takes more than ten years for a drug candidate to complete the Phase 1, Phase 2, and Phase 3 clinical trial processes usually required for drug approval. While some drugs can be approved based on Phase 2 data alone through an accelerated approval process, for the purposes of this example, we will consider the more typical path to approval.

The first category to consider is **time**. At TRACON, our experience has been that we can complete each phase of clinical trials one year faster compared to the typical clinical CRO timeline. This relates to the fact that we are financially motivated to do so, while clinical CROs are not. Another factor underpinning this accelerated developmental timeframe is the fact that we have "go-to" sites—more than sixty in the United States and many in Europe—that we know are high performers based on our past experiences with them, which are staffed by people with whom we have previously negotiated acceptable clinical trial agreements and budgets that greatly reduce study start-up timelines.

As noted in Figure 7, we not only "talk the talk" but also "walk the walk," as evidenced during our collaboration with Janssen, the oncology division of Johnson & Johnson. In that example, the TRACON PDP allowed for filing an Investigational New Drug application and completion of Phase 1 and Phase 2 trial enrollment within three years of the license for a drug candidate that had not yet been dosed to patients. Contrasting this timeline with the typical five-year period that many companies take to file an Investigational New Drug application and complete Phase 1 and Phase 2 trials provides a benchmark for the speed of the TRACON PDP. The fact that we executed the Phase 1 and Phase 2 trials on a total budget of $5 million attests to the cost efficiency of the TRACON PDP.

TRACON'S RAPID EXECUTION AFTER JANSSEN DEAL	
Deal signing, IND filing and initial dosing in 9 months	
TRC253 license executed	Sep 2016
IND filed	Dec 2016
Site Initiation	Mar 2017
First Patient First Visit	May 2017
Completion of Phase 1	July 2018
Completion of Phase 2 Enrollment	Sep 2019

Figure 7: Key milestones in the development of a prostate cancer drug candidate in a collaboration between TRACON and JNJ

The second category is **cost**. As we've noted, the TRACON PDP typically allows trial execution at a cost of approximately $100,000 per patient, which is typically less than half the costs typically paid to CROs. However, to be conservative we modeled that TRACON costs are half those of a CRO on a per-patient basis. As we've discussed, these cost savings stem from the fact that we implement the clinical trial at our cost and dispense with the need to double pay a CRO as well as our own employees to police a CRO. Rather, we pay our own employees to implement and monitor the study.

The third category is **quality**. For the purposes of this exercise, we were conservative and assumed the risk of success at each phase of development was identical between trial execution using the TRACON PDP and a clinical CRO. However, at TRACON we believe our model of shared cost, risk, and profit provides incentives to maximize quality that aren't compatible with the fee-for-service plus guaranteed payment model of CROs. Because TRACON interacts directly with study site personnel, we are much more likely to be made aware of an issue that may mandate termination of the trial for a safety concern. While no one wants a drug candidate to fail, facing the facts as early as possible (i.e., failing fast) is valuable to a pharmaceutical company, as doing so minimizes the cash outlay and allows reallocation of precious resources to other more promising drug candidates.

We also are more likely to uncover an issue that, if solved early in the trial, can allow the drug candidate to stay on track. For example, the drug Inrebic that I profiled earlier resulted in fatal brain damage in some patients as a result of exacerbating a vitamin deficiency. This key fact wasn't clearly understood until the Phase 3 trial was completed. The toxicity may have been eliminated if the issue was thoroughly understood earlier in development and each patient simply took a multivitamin. Because of the delayed recognition of a fatal side effect, the end result was a delay in the approval of Inrebic by five years after positive Phase 3 data and a Black Box warning on the drug label for possible brain damage that is not a risk for competitor products.

Adhering to the highest possible standards in our trials makes business sense, but it is also a moral imperative, because we develop drug candidates for people with cancer, often the rarer cancers with fewer effective treatment options. We need to ensure we dose trial participants appropriately, accurately assess outcomes in those

> ADHERING TO THE HIGHEST POSSIBLE STANDARDS IN OUR TRIALS MAKES BUSINESS SENSE, BUT IT IS ALSO A MORAL IMPERATIVE, BECAUSE WE DEVELOP DRUG CANDIDATES FOR PEOPLE WITH CANCER, OFTEN THE RARER CANCERS WITH FEWER EFFECTIVE TREATMENT OPTIONS.

patients, and detect potential issues with respect to safety and deal with those promptly. We believe this is best done by communicating directly with the hospital staff who are dosing patients with our drug candidates as compared to interacting with sites through personnel employed by a clinical CRO.

How important are the effects of streamlined clinical trial execution on the value of a drug candidate? We modeled the risk-adjusted NPV of a drug candidate that was projected to generate $1 billion in peak sales that was ready to initiate a Phase 1 clinical trial and developed using the TRACON PDP compared to typical CRO-based development. We assumed that a Phase 3 trial would be needed for approval and that the TRACON PDP would complete each phase of clinical trial development one year faster than a CRO at half the total cost, with TRACON paying those reduced costs. We assessed the value of the drug candidate to our partner using the TRACON PDP and profit-share deal structure compared to the value of them employing a CRO and commercializing themselves in the United States. As noted in Figure 8, the time and cost savings expected with the TRACON PDP utilizing the profit-share model without factoring in the advantages of superior quality are sufficient to approximately double the risk-adjusted NPV of the drug candidate to our partner.

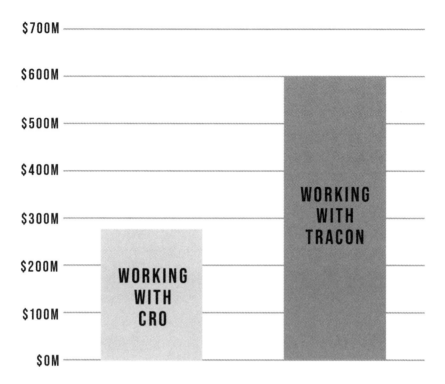

Figure 8: Net present value comparison to a partner of a drug developed over seven years at lower cost using the TRACON PDP versus ten years using a typical CRO

In summary, our partner reaps two means of increasing the retained value of their drug candidate by working with TRACON. First, we believe the time and cost efficiencies of our PDP can substantially increase the risk-adjusted NPV of a drug candidate. Second, our deal structure provides a further advantage to our partner compared to a Big Pharma license, as profit sharing enables our partner to retain an increased proportion of the increase in NPV absent the benefits of streamlined development enabled by the TRACON PDP (see Table 6). Accounting for both superior development efficiency and a more equitable sharing of economics provides significantly greater value to our partner.

Based on these advantages, TRACON has executed four deals in four years with pharmaceutical companies from the United States and China. Notably, in each deal we licensed a drug candidate or prehuman pipeline access *without* paying an up-front fee to our partner. We make the point to our partners that it is better for us to invest our capital into the shared product and fund clinical trials to speed development. In that way, we can achieve profitability as soon as possible, and our partners can reap a profit-share or equivalent large royalty in a rapid timeframe.

Chapter 8 ("Harnessing Global Innovation for US Patients") provides a case study of the promising drug candidate envafolimab that we have advanced into a registrational Phase 2 trial as an illustration of the value of shifting the typical drug-development paradigm. Instead of the "billion-dollar drug" paradigm, we are currently executing a drug-development program we believe will transform the care of sarcoma patients for less than $20 million in clinical trial costs leading to approval. This paradigm shift reflects multiple benefits—reducing time to market and reducing costs of clinical trial execution that generate higher NPV, streamlining development with potential approval based on Phase 2 data, and leveraging the investment of foreign-based partners with a dedicated drug-development program in their native country.

DOMESTIC AND FOREIGN-BASED PARTNERS

Our first corporate partner, Janssen, referred to working with TRACON as "strategic externalization." Our collaboration with them validated TRACON's PDP and helped us refine how we contracted with subsequent partners. Johnson & Johnson is one of the most forward-thinking Big Pharma companies in the world. The company's JLABS, for example, provides infrastructure throughout the United States and globally for innovative young companies to support discovery efforts. Ideally some of those products will advance to the point that Johnson & Johnson may want to license them for development and commercialization.

In 2016 Janssen partnered with TRACON rather than employing a clinical CRO to develop two oncology assets discovered in their laboratories. Our partnership with Janssen was one of the first of its kind, and it was notable in multiple ways. We received a $5 million investment from Johnson & Johnson, which covered the costs of conducting both Phase 1 and Phase 2 clinical trials for a drug candidate intended to treat patients who were resistant to approved drugs to treat prostate cancer. The drug candidate had been designed by Janssen to specifically reverse drug resistance in prostate cancer patients with a specific gene mutation. In return for the $5 million investment, we filed a dossier with the FDA to initiate the clinical trial process and subsequently completed Phase 1 and Phase 2 clinical trials.

While $5 million would not be sufficient capital to advance a prehuman drug candidate through Phase 1 and Phase 2 clinical trials using a traditional CRO-based model, it was sufficient using the TRACON PDP. The deal terms allowed Janssen to reacquire the lead asset in return for significant financial compensation to TRACON following completion of the clinical trials. The deal worked for both parties because TRACON was incentivized to deliver accurate data quickly and to Janssen's high standards to positively influence the reacquisition of the drug candidate, while at the same time encouraging us to be cost conscious. While Janssen ultimately did not opt to reacquire the asset due to the very small size of the population that would potentially benefit from the drug candidate, the deal with Janssen was important to TRACON in that it further validated the efficiency and value of our PDP.

Clearly, Janssen did not require TRACON's commercialization expertise, given they are part of Johnson & Johnson, the largest pharmaceutical company in the world as of 2021. Our optimal deal structure with our partners, most of whom are foreign-based drug development firms, includes securing US commercial rights, and one of the attractions of the TRACON PDP for foreign-based partners relates to the fact that they typically lack US development *and* commercialization expertise.

All the companies we subsequently partnered with after 2016 are based in China. We actively seek out companies possessing innovative drug candidates with first-in-class or best-in-class product attributes for which we envision a development pathway that will enable commercialization in a manner that can potentially transform medical care for specific unmet need patient populations in the United States. This is in contrast to a CRO, which may be indifferent to whether the drug candidate is first-in-class, best-in-class, or a me-too drug that will contribute very little to advancing the care of patients, since they earn the same profit from conducting a clinical trial or an innovative therapy as they do on a me-too drug.

Once we've identified a promising asset, we determine whether we can form a productive working partnership with the corporate management team of the innovating company. We carefully vet our partners to ensure they hold similar viewpoints regarding regulatory compliance and quality assurance and understand the regulatory hurdles that will need to be overcome to secure a drug approval.

In the case of a company with an antibody therapy, we also carefully perform diligence to ensure that suitable manufacturing capabilities are in place to provide for commercial scale production that will be necessary following drug approval. To date, our foreign-based partners are led by experienced management teams with significant experience at US pharmaceutical companies or the FDA. For example, one of our corporate partners is led by a former FDA reviewer.

Our goal in working with foreign-based pharmaceutical companies is to bring international innovation to US patients. Pharmaceutical innovation occurs throughout the world. And TRACON provides a solution by acting as a conduit between innovators and the US patient population. Our business development team uncovers innovation wherever it occurs in the world and then filters those innovations (or drug candidates) into a streamlined developmental and approval process, allowing for potentially transformative drugs to reach patients and deliver profits to innovators quickly and equitably through commercialization in the world's largest pharmaceutical market.

Our PDP and deal structure allows us to accomplish all that at a remarkably low investment. If a drug is discovered and developed (as well as manufactured) in China, we don't need to replicate that process in the United States. "All" we need to do on behalf of our partners and ourselves is prove the drug is safe and effective in US patients, which we feel we can do faster, better, and more cost effectively than any other option in the world, and then commercialize the drug following FDA approval.

The TRACON PDP is a revolutionary drug development and commercialization concept that offers benefits for multiple stakeholders in the biopharmaceutical ecosystem, including patient populations with unmet needs. When a drug candidate can be potentially approved for $20 million, more parties are willing and able to study innovative new drug candidates. Failures happen, but it's far easier to stomach a $20 million investment compared to the $2 billion investment typically required to approve each new drug.

CHAPTER EIGHT

HARNESSING GLOBAL INNOVATION FOR US PATIENTS

While the United States has been and continues to be the primary region driving drug-development innovation, other areas also contribute. Europe, Japan, and now China each have numerous innovative biopharmaceutical firms. For example, the ICI class of drugs (e.g., Keytruda®, marketed by Merck, and Opdivo®, marketed by Bristol Myers Squibb), which activates a patient's own immune system to attack his or her cancer, was developed based in part on academic research from Kyoto University in Japan. The patents from Kyoto University were licensed to the Japanese company Ono Pharmaceuticals and then subsequently licensed from Ono to Bristol Myers Squibb. The ICI class of drugs has revolutionized cancer care and is projected by *Barron's* to become the largest class of drugs (in terms of sales revenue) in the world within the next five years.[8]

International drug innovation is not limited to Japan. China, which initially became a major part of the drug-development landscape thanks to its large population, allowing for rapid recruitment into clinical trials, subsequently became a hotbed for companies adept at manufacturing small-molecule drugs and then antibody drugs and now has multiple companies focused on drug discovery.

BEST-IN-CLASS OPPORTUNITIES FOR UNMET NEED PATIENT POPULATIONS

As we've discussed, a potential first-in-class drug, one with an unproven mechanism for treatment, presents a well-documented high risk of failure for approval and eventual commercialization. It's certainly easier to develop a best-in-class drug that has a proven mechanism of action and focus the development process on addressing a patient population with an unmet need.

This was exactly the opportunity we perceived based on data presented in 2019 at the American Society of Clinical Oncology (ASCO) conference, the largest cancer conference in the world. Data were presented indicating that ICIs were active in certain soft tissue cancers, also known as sarcomas. Sarcomas are tumors of the muscles,

---

8   Josh Nathan-Kazis, "Merck's Keytruda Will Become Best-Selling Drug Worldwide, Research Group Says," *Barron's*, October 4, 2019, https://www.barrons.com/articles/mercks-keytruda-will-become-best-selling-drug-worldwide-research-group-says-51570204530.

bones, fat, and joints and are an aggressive tumor type without highly effective treatment options.

ICIs have now been approved to treat more than ten tumor types, and Keytruda is the current market leader in the class. The development path of Keytruda—a drug with applications in many cancer types—is similar to the one taken by IDEC and Genentech with Rituxan. Recall that Rituxan was originally approved to treat a rare subtype of non-Hodgkin's lymphoma, and clinical trials following initial approval were the basis for approvals in other cancer types as well as in rheumatoid arthritis. This type of "label expansion" is the ultimate goal of any transformational drug—to prove a novel mechanism can subsequently serve as a foundation to effectively treat multiple diseases. In the case of Keytruda, that label expansion has been rapid and thorough, as the drug label expanded from the initial approval in 2014 for treating patients with advanced melanoma to authorization of its use in more than ten cancer types by 2020 (Table 7). As a result of this expansion program, Keytruda is on track to become the top-selling drug in the world by 2025.

YEAR	INDICATION
2014	Advanced melanoma (initial approval)
2015	Advanced lung cancer
2015	Advanced melanoma (expanded indication)
2016	Head and neck cancer
2016	First-line lung cancer
2017	Classic Hodgkin's lymphoma
2017	First-line nonsquamous lung cancer with chemotherapy

YEAR	INDICATION
2017	Advanced bladder cancer
2017	Cancer with a certain genetic mutation
2017	Stomach cancer
2018	Cervical cancer
2018	B-cell lymphoma of the chest
2018	Lung cancer (expanded label)
2018	First-line squamous lung cancer with chemotherapy
2018	Liver cancer
2018	Merkel cell cancer
2019	Melanoma following surgery
2019	First-line lung cancer
2019	First-line kidney cancer
2019	First-line head and neck cancer
2019	Small-cell lung cancer
2019	Esophageal cancer
2019	Uterine cancer
2020	Early-stage bladder cancer
2020	Cancer with certain genetic mutation

YEAR	INDICATION
2020	Skin cancer
2020	First-line colon cancer with a certain genetic mutation

Table 7: Keytruda label expansion

ICIs have demonstrated remarkable activity, especially in lung cancer, melanoma and kidney cancer patients, where tumors tend to trigger a robust immune response. ICIs, such as Keytruda or Opdivo, essentially wake up a patient's immune system to attack a tumor. Tumors more likely to respond to ICIs have a high mutational burden, which means they display new tumor-associated proteins on their surface that can be recognized by the immune system as abnormal, which promotes an immune response. A good example of this process is lung cancer caused by smoking. Smoking carcinogens cause multiple mutations that engage immune cells that can attack the mutated proteins on the cancer cell surface. Unfortunately, this triggered immune response isn't active because the immune cells are held in check by the PD-1/PD-L1 checkpoint. ICIs like Keytruda and Opdivo override the PD-1/PD-L1 checkpoint (i.e., release the brake) and thereby allow the immune cells to attack the cancer (see Figure 9). However, many other cancers don't stimulate an immune response, in which case there is no checkpoint to override to enable immune cells to attack the cancer. Sarcomas generally fall into this latter category.

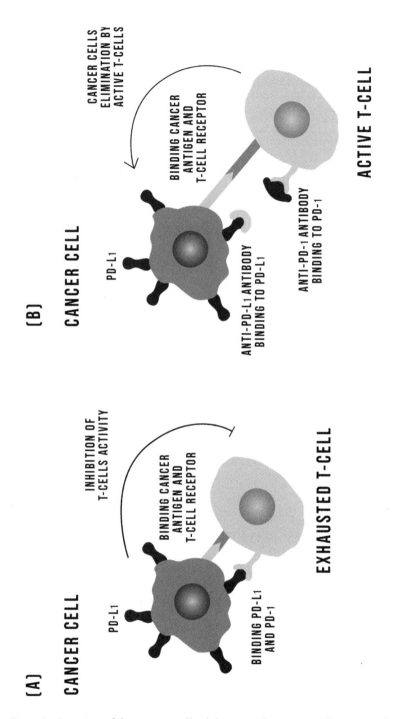

Figure 9: Overview of the immune cell inhibition and activation of immune cells through PD-1 or PD-L1 checkpoint inhibition

Importantly, the ASCO 2019 data indicated that certain sarcomas, subtypes known as undifferentiated pleomorphic sarcoma (UPS) and myxofibrosarcoma (MFS), were responsive to ICIs. In fact, Keytruda, when studied by a consortium of sarcoma key opinion leaders, demonstrated a remarkable response rate in a trial of patients with UPS and MFS who had progressed following chemotherapy (i.e., were refractory to chemotherapy; in cancer treatment terminology, "refractory" means a cancer that becomes resistant to treatment over time).

The Keytruda response rate compared quite favorably to the activity of the only approved drug for refractory UPS and MFS patients, Votrient®, a drug marketed by GlaxoSmithKline. The contrast was stark: Keytruda demonstrated a 23 percent response rate when used off-label in a trial of UPS or MFS patients whose cancers were refractory to chemotherapy, while Votrient offered a dismal 4 percent response rate. However, take a moment to look back at Table 7 charting Keytruda's long list of label expansion approvals and note what isn't on there: an approved indication use for treating refractory UPS or MFS patients.

Surprisingly, despite the impressive data, no pharmaceutical company was pursuing a clinical trial seeking to approve an ICI for these patients. By now we know the likely economic reason why: sarcomas—of which there are more than seventy subtypes—make up a small subset of cancer patients. There are fewer than five thousand cases of UPS and MFS annually in the United States, and a drug marketed solely for those patients would not be expected to generate revenue on par with an ICI marketed in a large indication.

But an initial market forecasted to generate $200 million in annual sales, using parity pricing to ICIs approved in large market indications, does make economic sense if drug development can be achieved at a far lower cost than that dictated by the usual $2 billion drug development paradigm. Using the TRACON PDP, we believed we could meet the unmet need of refractory UPS and MFS patients in need of a much better treatment option while also generating a substantial return on our investment.

> **USING THE TRACON PDP, WE BELIEVED WE COULD MEET THE UNMET NEED OF REFRACTORY UPS AND MFS PATIENTS IN NEED OF A MUCH BETTER TREATMENT OPTION WHILE ALSO GENERATING A SUBSTANTIAL RETURN ON OUR INVESTMENT.**

We employed the key principles that we have reviewed in this book and initiated a search for an ICI that we could license, develop, and commercialize in the unmet need population of US patients with refractory UPS and MFS. Our focus was China because we knew multiple companies there were developing ICIs targeting the PD-L1 pathway. Mark Wiggins and I visited China six times in 2019 with the goal of assessing every ICI that was being studied in clinical trials there. While we met with many companies that possessed an ICI that appeared to perform similarly to ICIs already marketed in the United States, most were administered in the same way, intravenously—through a vein, which requires an extended patient visit to an infusion center. In fact, all currently approved ICIs in the United States, including Keytruda and Opdivo, are given by intravenous administration.

We hoped there was a better way and conducted an intensive search for a drug candidate that was potentially best in class, by leveraging a proven mechanism of action while offering a significant benefit compared to existing treatments. We found what we were looking for after meeting two Chinese companies, 3D Medicines and Alphamab Oncology: a *subcutaneously administered* ICI that appears to be as active as market leaders Keytruda and Opdivo. We signed the license agreement with Alphamab Oncology and 3D Medicines in Shanghai on December 20, 2019, and returned to the United States just prior to the initial reports of COVID-19 in Wuhan, China.

There are a host of benefits to injecting a drug subcutaneously. Envafolimab, the PD-L1 antibody we licensed, is dosed in fewer than thirty seconds. Contrast that with the procedures a patient must endure when treated with an intravenous ICI: in that case, they need to book an appointment in the infusion center; typically receive premedication; have an intravenous catheter inserted through the skin into a vein that presents a risk of infection; wait at least thirty minutes for the infusion to be completed; risk an infusion reaction (which ranges in severity from shaking chills, nausea, and vomiting to wheezing and difficulty breathing that can occasionally be fatal), which may require further medical treatment; and then visit their physician in the clinic. Likely the morning, afternoon, or perhaps the entire day has been spent by an already exhausted patient to receive their infusion.

In contrast, envafolimab is dosed like a flu shot, and the patient can simply be treated in the clinic or physician's office. Following their physician visit, a nurse can administer envafolimab in the exam room in thirty seconds under the skin of the upper arm. Following the subcutaneous injection, the patient can go home soon thereafter due to the absence of the risk of an infusion reaction. The benefits of subcutaneous injection extend beyond patient and physician convenience. The simple method of administration in a clinic saves money for the healthcare system as well, compared to the rigmarole of prolonged intravenous dosing done in an infusion center under the supervision of a nurse or physician.

The subcutaneous route of administration achieves low variability in serum levels without the high peak concentrations seen with intravenously administered ICIs. This may confer a safety advantage for envafolimab. While this is yet to be proven, to date, envafolimab appears to be less likely to cause the side effects of inflammation of the large intestine and lungs compared to intravenously administered ICIs.

The envafolimab advantages suggest a truly revolutionary new angle for the treatment of sarcomas with an ICI. If successful, our ongoing development of envafolimab to treat specific refractory sarcomas will not only address an unmet need of this patient population but also potentially advance the entire field of cancer treatment by demonstrating the benefits of a more convenient therapy. And remember, UPS and MFS sarcomas are only two of the more than seventy different subtypes of sarcomas.

We are working with 3D Medicines and Alphamab to marshal our drug-development resources as part of our global strategy to approve and commercialize envafolimab as quickly as possible in the United States and worldwide. Envafolimab is an advanced drug candidate, as our partners completed the animal testing and other laboratory studies required to prove the mechanism of action of the drug and completed dosing Phase 1, Phase 2, and Phase 3 trials in China as well as Phase 1 trials in Japan and the United States. Our partners have submitted envafolimab for approval in China for patients with colon cancer and other cancers based on Phase 2 clinical trial data demonstrating a 32 percent response rate. Because our partners have also tested envafolimab in US cancer patients in a Phase 1 trial, the FDA has reviewed the manufacturing and animal safety testing documents and deemed them adequate to support continued testing envafolimab in US cancer patients in late-stage clinical trials.

Following our license for rights to develop and commercialize envafolimab for sarcoma treatment in North America, we designed the registrational ENVASARC Phase 2 trial as two trials in one. One goal, based on data from sarcoma patients treated with single-agent ICIs with a similar mechanism of action to envafolimab, is to achieve approximately a 15 percent response rate with envafolimab as single agent. In addition, we are studying envafolimab in combination with Yervoy®, an antibody that inhibits a second immune checkpoint called CTLA-4, with the expectation that we can increase the response rate in refractory UPS and MFS patients to 30 percent. This response rate target is based on data presented at ASCO 2020, demonstrating the combination of dual checkpoint inhibition using a PD-1 inhibitor combined with Yervoy resulted in a response rate of 29 percent in patients with refractory UPS.

Demonstrating that the response rate of envafolimab as a single agent and also in combination with Yervoy is superior to the 4 percent response rate of the only approved drug currently available for these patients (Votrient) may allow envafolimab to be approved as a single agent and in combination with Yervoy. While we expect that envafolimab would be used most commonly with Yervoy following its potential approval based on offering the prospects for a superior response rate, we also like the idea of providing single-agent envafolimab to patients who may be unable to tolerate the higher rate of side effects that may result from combining two ICIs. Notably, Bristol Myers Squibb executed a similar dual cohort clinical trial strategy for Opdivo in colorectal cancer and liver cancer that resulted in Opdivo's approval as both a single agent and in combination with Yervoy, with both approvals based on response rate.

TRACON announced the dosing of the first patient in the ENVASARC trial on December 10, 2020, in a press release that described the trial and our goals. We noted that we achieved the important milestone of initiating a registration trial less than one year of executing the license for the rights to envafolimab. We further noted in the release that we looked forward to interim top-line data from the study in 2021.

The key elements of the ENVASARC trial include the following:
- A multicenter, open-label, randomized, noncomparative, parallel cohort study at approximately twenty-five top cancer centers in the United States.
- Eligible patients will have undifferentiated pleomorphic sarcoma (UPS) or myxofibrosarcoma (MFS) and will have received one or two prior cancer therapies, but no prior ICI therapy.
- Planned total enrollment of 160 patients, with 80 patients enrolled into cohort A of treatment with single-agent envafolimab and 80 patients enrolled in cohort B of treatment with envafolimab and Yervoy.
- Primary end point of objective response rate with duration of response a key secondary end point.
- Open-label format with blinded independent central review of efficacy end point data.

As you will appreciate, we are adopting a strategy for the approval of envafolimab that is very similar to the one we adopted for Sutent when I worked at Pfizer. Namely, we are proposing to secure initial FDA approval based on response rate data for patients for whom approved therapies (Intron A in the case of kidney cancer; Votrient in the case of UPS and MFS) are not highly effective and can be quite toxic.

The FDA supports clinical trial designs that use response rate as the basis for accelerated drug approval for therapeutics that address patient populations with significant unmet needs (see chapter 6, "Understanding and Navigating the FDA"). In fact, three ICIs have been approved based on Phase 2 trials demonstrating response rates of 12 to 15 percent: Keytruda was approved to treat refractory stomach cancer based on a response rate of 13 percent, Tecentriq® was approved to treat refractory bladder cancer based on a response rate of 15 percent, and Opdivo was approved to treat a refractory form of lung cancer based on a response rate of 12 percent. Given the 4 percent response rate of Votrient, the only currently approved therapy for refractory UPS and MFS, we believe envafolimab combined with Yervoy could provide a transformative new standard of care for sarcoma patients.

At the same time, we're assessing biomarkers (e.g., unique receptors or unique genetic mutations in cancer cells) that may identify patients who respond particularly well to treatment with envafolimab. The incorporation of a biomarkers-based selection of patients most likely to respond to a drug candidate can be the difference between the success and failure of a clinical trial (see Figure 10).[9] Identification of a biomarker that predicts response to a cancer therapy allows the segregation of patients who are likely to respond to treatment from those who are unlikely to respond to treatment. For example, we've discussed genetic driver mutations that are highly predictive biomarkers of response (e.g., the JAK2 mutation in myelofibrosis or the BCR-*Abl* mutation in CML).

> **THE INCORPORATION OF A BIOMARKERS-BASED SELECTION OF PATIENTS MOST LIKELY TO RESPOND TO A DRUG CANDIDATE CAN BE THE DIFFERENCE BETWEEN THE SUCCESS AND FAILURE OF A CLINICAL TRIAL.**

Protein biomarkers can predict a greater likelihood of response to ICIs. For example, in some cases the protein biomarker PD-L1 expressed on cancers predicts a higher likelihood of response to an ICI like Keytruda or Opdivo. Merck, which markets Keytruda, used this

---

9   Data for Figure 10 drawn from David W. Thomas et al., "Clinical Development Success Rates 2006-2015," jointly published by BIO, Biomedtracker, and Amplion, June 2016.

biomarker to selectively enroll only patients with PD-L1-expressing cancer in a trial of patients with newly diagnosed lung cancer that demonstrated superior activity of Keytruda to chemotherapy. If Merck had enrolled patients regardless of PD-L1 expression, the trial may have failed to prove the benefits of Keytruda, because the treatment effect could have been diluted by lack of activity in patients whose tumors did not express PD-L1.

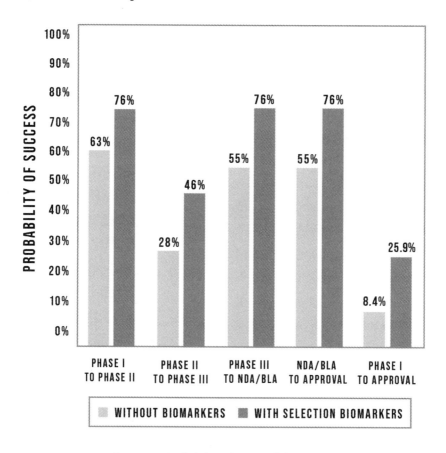

Figure 10: Probability of success of clinical trials with or without selection biomarkers (Source: BIO)

Concurrent with the expected initial approval of envafolimab to treat UPS and MFS as both a single agent and in combination with Yervoy, we plan to commence trials testing envafolimab's efficacy in other sarcoma subtypes, potentially using a biomarker to identify patients most likely to respond to treatment. In particular, we plan to combine envafolimab with chemotherapy in addition to the combination with the immunotherapeutic Yervoy being studied in the ENVASARC registration trial.

Generally, combining cancer drugs delays the development of resistance pathways and is a standard practice in the absence of major overlapping safety issues, which is why testing envafolimab with traditional chemotherapy is on our short-term agenda as part of our long-term potential label expansion strategy. Most newly diagnosed sarcoma patients initially receive chemotherapy. Therefore, if envafolimab can be given safely with chemotherapy, we may be able to treat sarcoma patients as soon as they are diagnosed rather than waiting for them to progress after initial treatment. Also, patients about to undergo a cancer resection who are at risk for cancer recurrence may benefit from preoperative therapy (also known as neoadjuvant therapy) as well as postoperative therapy (also known as adjuvant therapy) with envafolimab. Finally, patients with certain sarcoma types well served by existing treatments, like those with gastrointestinal stromal tumors (or GIST), may further benefit from the addition of envafolimab to standard-of-care medication.

Overall, our goal is to ensure that every sarcoma patient who would benefit from an ICI has access to envafolimab. Expanding the approved uses of envafolimab (i.e., label expansion) represents the natural evolution of envafolimab development in sarcoma. We're advancing this strategy by focusing initially on the highest unmet need patient populations where activity appears to be greatest and then continuing development following initial approval to substantially increase patient access and market potential.

THE ENVAFOLIMAB PARTNERSHIP AS A MEANS TO GLOBAL COMMERCIALIZATION

We are working in partnership with 3D Medicines and Alphamab Oncology to approve and commercialize envafolimab for sarcoma patients globally. TRACON is conducting and bearing the costs of envafolimab's registrational ENVASARC clinical trial in refractory sarcoma subtypes, which is designed to provide evidence of clinical benefit for approval in the United States. In the meantime, our corporate partners performed the necessary activities to submit envafolimab for approval in colon cancer and other cancer types in China in 2020. Approval in many other indications could then follow in the United States, China, and elsewhere.

Our partnership with 3D Medicines and Alphamab Oncology serves as an example of the value of expected reduced time to market through a customized partnership structure that utilizes the TRACON PDP and the discovery, manufacturing, and clinical developmental investments of a foreign-based partners. We estimate the budget to generate the clinical trial data needed to initially approve envafolimab in the United States to be less than $20 million—the majority of that budget being the cost of the ENVASARC trial implemented using the TRACON PDP. This cost is far less than that exemplified by the $2 billion drug paradigm that typifies the pharmaceutical industry. Part of that cost savings reflects the TRACON PDP advantages. However, our time and cost investments are also proportionally lower because all three companies leverage and benefit from each other's investment and expertise (Table 8). These efficiencies are important so we can focus on a smaller market indication and still expect a high return on investment based on our shared costs, TRACON's efficient PDP, and shared US commercialization rights. We expect that our partners will further benefit by being able to file for approval in China and elsewhere to treat sarcoma patients using the same data we expect to provide to the FDA for US approval.

- Discovery, animal testing, manufacturing, and late-stage clinical trial trials completed by our partners 3D Medicines and Alphamab to support approval in China in colon cancer and other cancer types.

- Envafolimab is a potential best-in-class drug (i.e., one with a validated mechanism of action that lowers the clinical risk of failure that is given by a more convenient method of administration).

- Envafolimab is being developed by TRACON to address a patient population with a high unmet need (refractory UPS and MFS), which allows for potential accelerated approval through Phase 2 trial data.

- Development is done using the TRACON PDP to permit rapid, low-cost, and high-quality clinical trial execution.

- TRACON commercial expertise will be utilized to market envafolimab to sarcoma physicians in the United States.

- The dossier submitted for approval to the US FDA is expected to allow for approval and subsequent commercialization throughout the world.

Table 8: Factors leveraged to develop and potentially commercialize envafolimab in sarcoma in the United States and then globally

Reducing time to approval and commercialization has another benefit: maximizing the patent life of the drug candidate. A drug patent generally lasts twenty years, and that clock starts ticking from the patent filing date, which is typically done at the start of the drug-development process. The faster a drug is commercialized, the more time it has to help patients and reward researchers and investors who were partners in the effort to bring the medication to market.

We believe that, if approved, the commercialization of envafolimab to the sarcoma community will be straightforward. In the cancer community, a small number of physicians specialized in treating a given cancer type typically influences the practice patterns of other practitioners. They also serve as key opinion leaders when communicating a drug's value to the FDA. Implementing the ENVASARC clinical trial at hospitals with key opinion leaders provides the experience during development that can be referenced and leveraged following approval during marketing of the drug. In a rare tumor type like sarcoma, the community is intimate, and we involve key opinion leaders in our trials to provide them with direct experience using envafolimab.

If envafolimab is approved in sarcoma, those physicians will have firsthand experience with the drug and can speak to its benefits when they talk with their patients and when they speak at conferences, where everyone is focused on identifying the highest standard of care for patients. Furthermore, these physicians are expected to prescribe the drug—as they will be highly incentivized to provide their patients a superior treatment with better safety and activity, as well as the convenience of subcutaneous injection.

This is also why securing widespread adoption of a new medication with a transformative efficacy and convenience profile to a relatively small patient population is easy to support and reinforce with a small group of sales representatives. The number of physicians prescribing envafolimab to sarcoma patients is expected to be concentrated, with the expectation that general medical oncologists in the community will follow the lead of the experts. That's very different from marketing a drug for a more common disease, such as diabetes or high blood pressure, where sales representatives must meet with thousands of physicians to convince them to use their product rather than one of several others available from competing pharmaceutical companies.

In comparison, envafolimab sales representatives are expected to easily expand their reach among the cancer care community through touting the benefits of envafolimab, aided by the lack of highly effective competing drugs in this space. And the success of a new treatment in the oncology space is something everyone wants to talk about as well as popularize. This topic is further explored in chapter 10 ("The Finer Points of Commercializing an Oncology Drug").

CHAPTER NINE

INTEGRATING TECHNOLOGY AS PART OF THE TRACON PDP

The value we offer to our partners isn't just a matter of our years of expertise, CRO independence, and equitable profit-share deal structure. We have also developed a superior drug-development infrastructure at TRACON. Our experiences at Big Pharma and biotechnology companies showed us that the technology used to manage clinical trial data and FDA reporting is inefficient due to a lack of integration between systems, which results in duplication of effort and ultimately contributes to the high cost of drug development.

What informed TRACON's decision to invest in a technology solution as the backbone of our PDP was Bonne Adams's realization crystallized and communicated to the executive team back in 2007—namely, that it would be far more cost effective to do the work of a CRO directly, rather than pay a CRO and suffer the consequences of a poorly aligned model of development. She leveraged that realization to optimize technology and develop in-house software solutions capable of retrieving information from multiple sources and collating data in an FDA-compliant manner. Her vision was an integrated platform with systems that talked to each other.

> IT WOULD BE FAR MORE COST EFFECTIVE TO DO THE WORK OF A CRO DIRECTLY, RATHER THAN PAY A CRO AND SUFFER THE CONSEQUENCES OF A POORLY ALIGNED MODEL OF DEVELOPMENT.

Bonne's extensive experience informed how a best-in-class drug-development software platform should operate, and we have used that experience to encourage and enable our vendors to create highly efficient integrated software solutions. The value of our efforts to TRACON and our development partners is significant, as we are ultimately able to develop more drugs with fewer personnel at high quality. Even more importantly, we create value for other companies in the biopharmaceutical ecosystem, who can also contract with our software vendors to benefit from the capabilities developed for TRACON, creating an industry-wide virtuous circle of efficiency and rigor. To understand how the integrated TRACON technology platform works, it's important to understand all the ways inefficient technology contributes to the high cost of the clinical drug-development process.

TECHNOLOGY

There are two primary players that collect data during the conduct of clinical trials: the pharmaceutical company, and the hospitals and clinics who treat patients. Many pharmaceutical companies have thousands of employees who use outdated software solutions that define processes and standard operating procedures utilized across huge global organizations. It is very difficult for these large firms to adopt the latest technology solutions in a timely manner, as they manage hundreds of ongoing clinical trials at various stages of development (and changing software in the middle of a clinical trial is unwise), and adopting new technology solutions means they must retrain thousands of employees on how to use them. In addition, from a technology perspective, pharmaceutical company departments such as pharmacovigilance (i.e., safety monitoring), clinical operations, clinical development, and data management largely operate software solutions that are not integrated, and the lack of data sharing therefore results in a significant duplication of effort.

Hospitals and clinics participating in clinical trials also use many different software platforms to collect and store patient information; some even still maintain paper charts. These software solutions are called electronic medical record (EMR) systems and have largely been inaccessible by pharmaceutical companies. What happens is this: the hospitals and clinics participating in clinical trials review data collected on patients participating in clinical trials from their EMR and then reenter those data into a database system used by the pharmaceutical companies.

The next big technological advancement underway in the pharmaceutical industry is the integration of hospital and clinic EMR systems with pharmaceutical company databases. This step will greatly reduce the duplication of effort and burden placed on hospitals and clinics participating in clinical trials because they will no longer have to enter the same information into multiple systems (their own, and that of the biopharmaceutical company conducting the clinical trial). It will also greatly reduce the duplication of effort on the part of the pharmaceutical companies, as they will only need to review information from a single database. Integration of EMR systems with pharmaceutical databases will also allow for information about a drug safety profile to be disseminated more quickly, as the delay between the recognition of the event and entry into the pharmaceutical company database will be eliminated.

Further inefficiencies exist within the Big Pharma firm or CRO implementing the clinical trial. Each of the entities contains several large departments: pharmacovigilance, which deals with safety issues; clinical operations and clinical development, which is responsible for overseeing the trials; biometrics, which is responsible for data management and statistics; chemistry manufacturing and controls, which is responsible for manufacturing research drugs; regulatory affairs, which is responsible for corresponding with FDA; and bioanalytical, which is responsible for patient sample testing and analysis. Many times, these departments use different software solutions, and many of the solutions don't talk to each other.

Let's take a hypothetical example of a single clinical trial patient to demonstrate the issues stemming from a lack of communication between software solutions. Prior to enrolling in a clinical trial, a patient has received a full medical workup, including labs, physical exam, baseline ECG, and other tests, and those data are housed in their doctor's office system. These data must be manually entered by the doctor's team into the CRO or pharmaceutical company data management system. Because these data are manually entered from one system into another, it must be checked to ensure there aren't any data entry errors; the checking is typically done by medically trained staff from the CRO or pharmaceutical company. The medically trained staff from the CRO or pharmaceutical company will request that the doctor's team correct errors; the doctor's team makes corrections, and then the CRO or pharmaceutical company will re-review the data to ensure accuracy. If the doctor's medical record system was integrated with the CRO or pharmaceutical company database so relevant data could be automatically transferred electronically, hundreds of work hours spent by both the hospital staff and the pharmaceutical company could be allocated elsewhere.

The process gets more complicated. Now imagine the hypothetical patient has a serious reaction to the investigational drug being studied in the clinical trial. The patient presents at the emergency room of a hospital near their house (not the hospital or clinic where they are being treated as part of the clinical trial). The patient is admitted to the hospital, has numerous tests during their hospital stay, and then recovers and is eventually released from the hospital. If the hospital where the patient was treated is not affiliated with the same medical system as their clinical trial doctor's system, then all of the results from the tests conducted during this hospital stay must be manually entered into the medical record system at the clinical trial site. This same information must also be entered into the CRO or pharmaceutical company data system.

It gets even more complicated. Within the pharmaceutical company, the pharmacovigilance department is responsible for collating serious side effects and ensuring that they are reported to the FDA. Typically, the pharmacovigilance department collects serious-side-effect data from the study in a database *separate* from the one maintained in the data management department. Therefore, the doctor's office must enter all the information received from the hospital that treated the patient in two different systems for the pharmaceutical company. In short, the doctor's office must rapidly collect information from the hospital where the patient was treated and enter that information into *three* different systems (their own EMR, the pharmaceutical company's data management system, and the pharmaceutical company's pharmacovigilance system) and are under pressure to do it very quickly, as this type of information must be reported to the FDA within one or two weeks when the event is unexpected and related to the study drug (see Figure 11).

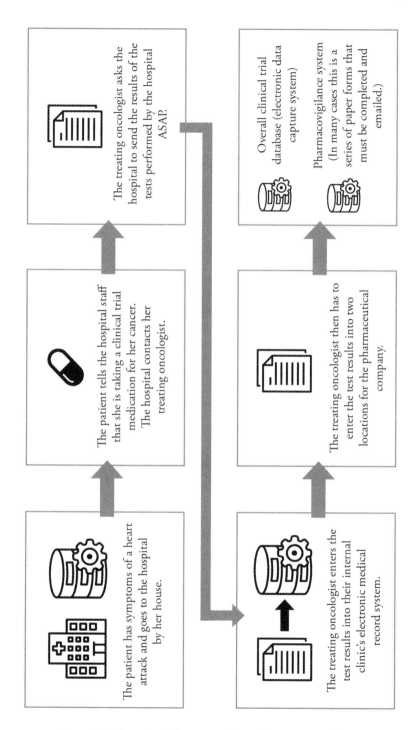

Figure 11: Schematic of the process of recording a serious adverse event in a clinical trial participant

Let's resume our patient's journey. Once the Big Pharma company has been notified of a serious side effect, multiple departments process the event. The pharmacovigilance department compiles all of the information about the case in a narrative format, the regulatory affairs department submits the information to the FDA, and the data management department maintains the information within the electronic database that houses all the patient's information. *However, none of the systems are integrated with each other.* The information required by each department often must be manually entered and compared to the same information entered into another system, whose profiles are shared archaically—via an emailed PDF or fax. Once reconciled in all the different departments at the Big Pharma firm or CRO, the information must then match and be formatted in the correct reporting manner to be shared with the FDA.

> OUR CLINICAL OPERATIONS AND BIOINFORMATIC DEPARTMENTS REPRESENT A SINGLE LOCATION FOR DATA COLLECTED AS PART OF A CLINICAL TRIAL THAT IS INTEGRATED WITH OUR OTHER PHARMACEUTICAL SYSTEMS TO MINIMIZE DUPLICATIVE EFFORT.

The result is a very high potential for error due to duplicative data entry, representing an extraordinary waste of time and money. Conversely, ensuring all systems talk to each other, are programmed to detect variances that require immediate attention or explanation, and collate data in the correct format required for reports submitted to the FDA and regulatory agencies in other countries vastly increases efficiency and safety. TRACON has integrated our internal systems and is working toward the direct integration of our systems with hospital EMR systems to eliminate this duplication of effort.

We have and continue to work with data management software vendors to create a far more effective system. Our clinical operations and biometric departments represent a single location for data collected as part of a clinical trial that is integrated with our other pharmaceutical systems to minimize duplicative effort. The system is compliant with the Code of Federal Regulations and has compliant audit trails. These audit trails provide transparent insight into when and how data has been updated or changed. That's an important consideration for the FDA, which demands the ability to audit any aspect of a clinical trial. TRACON systems interact according to the following diagram and can immediately generate narratives, reports, data tables, and data listings.

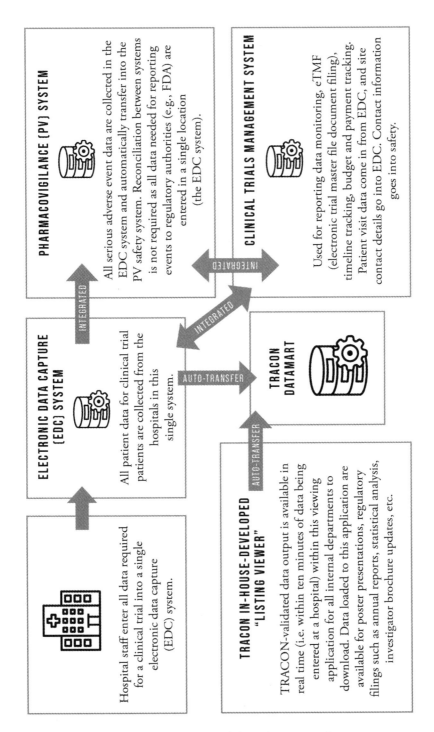

Figure 12: Schematic of an integrated clinical operations data management solution used at TRACON

We have worked extensively with our software vendors to integrate electronic data capture (EDC), safety, and clinical trial management solutions. These systems are able to automatically transfer data from a single source into the various formats needed by different internal departments, by our partners in development, and by regulatory authorities, all of which have discrete and different requirements for different types of data. These transfers are accomplished seamlessly and efficiently without duplication of effort. Audit trails are available for all inputted data, changes, exchanges, and outputs along the way. This software integration reflects the value of decades of experience and countless work hours of effort.

The clinics and hospitals we work with in our clinical trials also benefit from our software integration efforts, as they need only enter data collected from patients treated at their clinics enrolled in our clinical trials in a single place, our EDC system. All patient data are entered into the EDC system, including demographics, past medical history, and data collected as part of the ongoing clinical trial, such as drug dosing information, side effects, lab results, physical exam data, and vital signs (see Figure 12). Our next objective is integrating our EDC system with the patient's EMR so that information will not need to be entered from one system into another by hospital staff. This is an industry-wide aspiration and something TRACON is working to achieve in collaboration with multiple software vendors. However, EMR to EDC integration is complicated, as not all clinics or hospitals use the same EMR systems, and many clinics and hospitals still maintain paper charts. When hospitals and clinics can electronically transfer critical data from a clinical trial patient to the pharmaceutical company EDC system, there will be even greater improvement of efficiency and data quality.

Our safety reporting system is fully integrated with our EDC system. For example, the clinical trial site enters information about a serious side effect into the EDC system, which then automatically notifies a series of individuals at our company. Our development partners are also copied on these emails to allow for prompt automatic notification, regardless of the time of day. Data flows from the EDC in a single direction into the safety system. The TRACON safety team evaluates data that have been entered in the EDC and are then transferred electronically into the safety system, which automatically generates case narratives describing the event, houses source documents including hospital discharge records, and generates reports for the FDA and other regulatory authorities. When our safety team requires additional information about a case, they query the hospital via the EDC system so the site may enter or update the needed information in the EDC system. The updated information is then retransferred into the safety system. By having the data flow in a single direction and making the EDC system the master system for all data, the need for database reconciliation between the safety and EDC systems is eliminated. In addition, we maintain hospital records in a single location (the safety system document repository).

We have also licensed a Clinical Trial Management System (CTMS) from the same software vendor that supplies our EDC and safety solutions. The CTMS is used for maintaining data monitoring reports, maintaining key clinical trial site documents, and housing contact information, timeline tracking, and payment tracking. Patient visit information transfers over to CTMS from the EDC automatically, enabling the management of enrollment timelines, drug supply projections, budgets, and payments. Contact information stored in the CTMS flows into the EDC system and the safety system so that there is a single source of hospital and internal staff contact details. This also means that there is a single location to maintain site and patient numbering, allowing for accurate access of patient data within each component of the system.

We have also implemented our own internal data mart for housing raw data in a manner that is compliant with federal regulations. Data are transferred into our in-house Listing Viewer application, which allows review of each data point for each enrolled patient in a TRACON-sponsored clinical trial. TRACON-validated data output is available in real time (i.e., within ten minutes of data being entered at a hospital) within this viewing application for all internal departments. Data loaded to this application are available for poster presentations, regulatory filings such as annual reports to the FDA, statistical analyses, and investigator brochure updates. Finally, we also have plans to integrate quality and regulatory applications with our existing software solutions to make the TRACON in-house platform even more efficient.

A critical part of a trustworthy technology platform is quality assurance accomplished through validation. It has taken hundreds of work hours to create and validate a platform of software that accurately documents every single aspect of a drug trial in a way that is tracked and audited (in accordance with the Code of Federal Regulations). While the effort to accomplish an integrated and validated clinical operations and development system should not be underestimated, as Bonne points out, "We only had to do it once, and now that it is set up, there is no limit to the number of studies TRACON can support utilizing our integrated software solutions."

CHAPTER TEN

THE FINER POINTS OF COMMERCIALIZING AN ONCOLOGY DRUG

Most of this book has focused primarily on drug development, including understanding the FDA approval process. But then comes the all-important commercialization component. Commercializing a drug for a specialty audience of a distinct subset of physicians, such as oncologists, is very different from what people think of when we talk about commercializing a drug for a general practitioner audience. When you think about the number of doctors in America who prescribe medications for hypertension of diabetes, that's in the tens of thousands. Reaching that group, especially given the many similar drug choices for common diseases like hypertension, is a huge undertaking that requires thousands of sales representatives constantly fighting for mindshare against just-as-good competitor products.

That's why Big Pharma companies have dedicated sales forces consisting of thousands of representatives to market to that audience. This could be viewed as another example of a waste of resources in chasing a piece of the profit pie rather than dedicating money to innovative drug development, as Big Pharma companies underwrite thousands of sales representatives to convince doctors to prescribe their drugs for common diseases like hypertension or diabetes, rather than a competitor product that may have a very similar safety and efficacy profile.

Conversely, to market a cancer drug for an unmet need patient population, you initially need a total sales team of sixty or fewer people. Even a massive broadening of that team would, at the most, double it to 120 people. And you can easily do that in a small company, like one the size of TRACON. The group that facilitates commercialization isn't only sales representatives; it also includes medical science liaisons who directly interact with physicians to answer scientific questions. In addition, the commercialization team includes a small medical affairs group, people who will talk and listen to physicians to share insights and interests in additional trials that may show additional benefits of a new drug. When you have such a small group (60 to 120 people to reach and interact with an entire specialized community), you can manage it extremely thoughtfully and well. Everyone is engaged with the community and with each other, and it is easy to share information and feedback.

With envafolimab, for example, our commercialization team is expected to primarily interact with physicians who treat sarcoma patients, as they are the ones expected to prescribe our drug. Many of the physicians who prescribe a new cancer drug, especially one targeted to an unmet need patient population, are often involved in the initial clinical trials. It makes sense when you think about it. The sarcoma patient population is relatively small, compared to other cancers. Physicians treating sarcoma patients, especially those with refractory sarcomas, are as motivated as their patients are to study a treatment that offers a potential benefit compared to existing treatments. As a result, they're already highly incentivized to enroll their patients in promising clinical trials.

Those physicians gain direct experience with a drug candidate through treating their patients as part of a clinical trial. Thus, when the drug advances into the FDA approval process, those physicians also function as informed key opinion leaders, to whom the FDA listens when, for example, the agency is making a decision to grant accelerated approval (based on Phase 2 trials) or full approval (based on randomized Phase 3 trials). Regardless of which type of approval the drug receives, those physicians also become a reference point for other physicians to spread the word about a better treatment option for their cancer patients, which contributes to driving the adoption of that new drug. These doctors are also invited to speak at conferences, where they are eager to share new innovations in the field that raise the standard of care for patients. Once doctors like these are thoroughly versed in the benefits of a successful new cancer drug for an unmet need patient population, a company like ours needs very few actual sales representatives to facilitate widespread adoption because, by that point, the sales reps are simply reinforcing what the sarcoma physician community is already propagating among themselves.

At this point (early 2021), envafolimab is being studied in the registrational ENVASARC trial at the top sarcoma centers in the United

States. Nearly every American key opinion leader in the sarcoma community is aware that we're conducting a trial in sarcoma patients with limited treatment options using a subcutaneously administered ICI, and they fully understand the convenience of subcutaneous injection compared to intravenous administration.

As you can appreciate, that type of awareness and engagement is a vastly different and far more economical proposition than hiring thousands of sales representatives for a commercialization initiative of the type that is typically employed by a Big Pharma company marketing yet another drug to a general practitioner population already swamped with competing drugs. In the oncology space, especially the unmet needs oncology space, physicians are eager to hear about and/or seek out information about new and potentially transformative treatments, simply because such treatments are so rare. This contrasts with general practitioners, who deal with a daily barrage of incentives to switch from prescribing one type of hypertension drug to another, without any truly compelling medical reason for doing so.

If envafolimab continues to perform as expected in our current and future Phase 2 and Phase 3 trials, we won't have to fight for mindshare among prescribing physicians. They'll already be well aware of the drug and its benefits, and eager to proselytize it among their colleagues. Our sales support won't have to blanket the country to make an impact among our prospective group of prescribers; they'll simply have to work closely, intelligently, and responsively with a relatively small group of physicians who are eager to hear from them. It's a win on every level: for patients, oncology physicians, TRACON, and our partners.

What we're proposing to do with our sales force/commercialization team once envafolimab is at a place where it can be fully commercialized in the United States is not necessarily innovative. It's simply a factor of smartly leveraging our resources to support an efficient business development initiative (i.e., effective commercialization). We've gotten in the habit of seeing where bloat and inefficiencies exist in the pharmaceutical ecosystem, and the commercialization aspect is no different in that regard than any other part of the process. It isn't necessarily new for a company to leverage the interest and participation of specialty oncology physicians in drug trials; we're just mapping our planned commercial efforts in a way that acknowledges and reinforces the value those physicians provide in helping spread awareness and adoption of a new drug.

COMPENDIA LISTINGS AND KEY OPINION RESEARCHERS

There is an important parallel path for the adoption of drugs approved to treat unmet patient populations. It involves a consortium of doctors and other medical professionals who contribute data, insight, and awareness through trials they conduct themselves, independent of a pharmaceutical company, using a drug supplied to them by the pharmaceutical company. The information collected by these professionals serves numerous valuable purposes. It provides doctors who are treating rare cancers with information on the activity of new types of treatments, where they are simultaneously being educated about a new treatment and providing education through their use of that treatment. Their data can be used to support the FDA's requirements for further information to support the approval of a drug to treat an unmet disease. Furthermore, the information these doctors and other medical professionals provide—especially when it involves using a new drug to treat other types of cancers—can be used to support insurance reimbursement claims. For example, if enough doctors use a drug off-label to successfully treat a cancer, that practice can become part of the compendia. In turn, compendia data support movements on behalf of insurance companies to expand reimbursement for the drug's use in indications beyond those approved by the FDA. In oncology, the National Comprehensive Cancer Network (NCCN) compendia are some of the most valuable, in that drugs listed by the NCCN for unapproved indications based on scientific evidence are frequently reimbursed by Medicare and other large third-party payers when used off-label.

SUMMARY

A Pharmaceutical Paradigm Shift as an Antidote to the Billion-Dollar Drug Problem

TRACON TODAY: FOCUS ON UNMET NEEDS

We are particularly focused on sarcoma patients because ICIs have demonstrated activity in this cancer type, yet none are approved for use in this rare tumor despite the lack of other effective treatments. Envafolimab represents a means to meet this unmet need using a potential best-in-class ICI by virtue of its fast and convenient method of rapidly delivered subcutaneous administration.

Successfully commercializing a best-in-class drug in sarcoma will allow TRACON to transition from a developmental stage company, which we are now, into a commercial company, which is our goal in 2023. We then expect to replicate the process of harnessing global innovation to meet the needs of cancer patients without effective treatment options over and over again, either testing drug candidates in combination with envafolimab in sarcoma or new drug candidates in additional patient populations with high unmet needs.

To meet that goal, we expect to repeatedly employ the CRO-independent TRACON PDP to select promising drug candidates, conduct clinical trials, obtain initial approval, commercialize, and then expand into new indications. Our long-term vision is the reason we've spent years investing in a scalable model for rapid and high-quality drug development.

We believe the TRACON PDP is a long-term solution that can be applied to multiple drug candidates ready for clinical testing, whether that's initial Phase 1 trials or Phase 2 or Phase 3 trials. Our platform and profit-share deal structure are ideally suited to foreign-based companies that do not have a solution for developing or commercializing their drug candidates in the United States without ceding much of their value.

SHIFTING THE PHARMACEUTICAL PARADIGM: AN ALIGNMENT OF INTERESTS—THE TRACON APPROACH

To revisit the question I posed in the introduction: Can the $2 billion price tag per successful drug be reduced significantly, allowing for a far more efficient process—one that offers greater incentive for innovation and focus on rarer patient populations with unmet needs?

By now, I hope you agree that the answer is yes.

As you've seen, the costs associated with testing a drug are significant, due in no small part to the outsized role of CROs. Well-run CROs absolutely have their place in a global pharmaceutical ecosystem. But they are by no means a one-size-fits-all strategy. Small drug developers beholden to the misaligned CRO business model risk or forfeit time, money, and even the ability to stay in business long enough to develop new drugs.

It is possible to develop an in-house model that leverages the best that CROs offer, by bringing best-in-class oversight and technology in-house to prioritize rapidity and reduce cost without sacrificing rigor. TRACON's unique PDP originated through the amalgamation

SUMMARY

of viewpoints from experts with decades of experience in the biopharmaceutical sector. It maximizes the value of drug candidates for our partners through cost-effective development and through a superior profit-share approach, as compared to the typical Big Pharma deal structure. Widespread adoption of our model could increase the successful development of first-in-class or best-in-class drugs at a far lower investment than the norm, which in turn could negate the need for premium drug prices to generate a significant return on investment for drugs dedicated for the treatment of rare diseases.

We use our combined expertise to harness global innovation to identify promising drug candidates, evaluate potential partners who are aligned with our quality standards, and invest in aiding software companies to develop capabilities that streamline data management and reporting that support comprehensive, thoughtful, and effective paths to approval. As the examples in this book show, today an innovative drug company anywhere in the world need *not* take the typical path of licensing their drug candidate to a Big Pharma company and ceding the majority of its value. Aligned development, such as through the TRACON profit-share deal structure, best serves the interests of partners and patients alike.

We encourage companies within the pharmaceutical ecosystem to take a hard look at what they may perceive to be necessary and unchangeable practices. To reconsider the value of investing vast amounts of resources into me-too-with-a-twist drugs for patients already well served by similarly approved drugs. To contemplate investing in technology systems to facilitate more cost-effective drug development that enables a return on investment for drug candidates dedicated to patients without effective therapies. While the TRACON PDP is proprietary, the technology systems we use are not.

This realization could not come a moment too soon. Ideally the lessons in the book illustrate that TRACON expects to approve a drug for one-hundredth of the commonly accepted cost of $2 billion. I hope this book encourages all participants in the drug discovery and development field dedicated to advancing patient care to reevaluate current paradigms in the pharmaceutical ecosystem and consider their opportunity for improving them to the benefit of all.

ACKNOWLEDGMENTS

This book wouldn't be possible without the efforts of three of TRACON's senior executives and coauthors of this book. The Product Development Platform was the brainchild of Bonne Adams, TRACON's Executive Vice President of Clinical Development and Clinical Operations. Necessity is the mother of invention, and early in our history Bonne internalized clinical operations at TRACON to reduce costs and optimize clinical trial quality, which was necessary for us to secure venture capital funding. She then upgraded the system to become a standard in our industry. In addition to her passion for cancer care, Bonne is committed to serving others and leads TRACON's charitable outreach, which funds organizations like Feeding San Diego. Mark Wiggins, our Chief Business Officer, leveraged Bonne's brainchild by using his years of experience to devise the profit share deal structure that creates alignment between TRACON and its corporate partners to allow for the harnessing of international innovation to benefit global populations. He also is an avid wine connoisseur and has a wickedly dry sense of humor. Scott Brown, TRACON's Chief Financial Officer, developed models to quantify the advantages of aligned drug development, validating a drug development paradigm that logically makes more sense. His talents extend far beyond mathematics, and he is a dedicated CrossFit athlete. Their passion, as well as the passion of each of my coworkers, inspires me daily. I am also indebted to the team at ForbesBooks, especially Deborah Roth, whose painstaking efforts working with me allowed for the preparation and execution of this work.

GLOSSARY

TERM	DEFINITION
AIDS	Acquired immunodeficiency syndrome; a chronic, potentially life-threatening condition caused by the human immunodeficiency virus (HIV).
Allopurinol	(Zyloprim®, Aloprim®) A drug that inhibits uric acid formation in the body that is the most commonly used medicine to treat gout. Developed by George Hitchings and Gertrude Elion, who were awarded the 1988 Nobel Prize in medicine for their work in developing allopurinol, azathioprine, and five other drugs.
Allos Therapeutics	Pharmaceutical company based in the United States that focuses on the development and commercialization of anticancer therapeutics. Markets the orphan drug Folotyn for the treatment of orphan drug indication of peripheral T cell lymphoma that costs more than $30,000 per month.
Antibodies	Natural proteins secreted by B-lymphocytes that bind to specific target molecules on the outer surface of cells with high precision, including those on cancer cells.

ADC	Antibody-drug conjugates; highly targeted biopharmaceutical drugs that combine monoclonal antibodies specific to surface antigens present on particular tumor cells with highly potent anticancer agents linked via a chemical linker. ADCs are a powerful class of therapeutic agents in oncology and hematology. Also referred to as immunotoxins.
ASCO	American Society of Clinical Oncology; professional organization representing physicians of all specialties who care for people with cancer. Holds the largest annual conference on cancer treatment.
AstraZeneca	British–Swedish multinational biopharmaceutical company. Purchased San Diego–based biotechnology company Ardea for $1.2 billion for the gout drug candidate Zurampic®. Subsequently sold Zurampic to Ironwood Pharmaceuticals in 2016.
Avastin®	(Generic: bevacizumab) An antibody drug marketed by Genentech that inhibits the growth of cancer blood vessels and cancer cells by binding and neutralizing vascular endothelial growth factor (VEGF). Avastin® is approved to treat common tumor types of cancer including colon and lung cancer in combination with chemotherapy.
B-cells	Lymphocytes that produce antibodies.
Bayer	One of the largest multinational pharmaceutical companies in the world and headquartered in Germany. Markets the drug Stivarga®, which was approved in 2012 for colon cancer patients, which works by a mechanism of action to similar to Sutent®.

GLOSSARY

BCR-*Abl* mutation	A specific DNA mutation caused by the fusion of two genes that causes chronic myelogenous leukemia (CML), which is effectively treated with Gleevec®.
Best in Class	A drug that improves upon a validated and approved first-in-class mechanism to potentially offer the "best" efficacy and safety profile for a specific indication.
Big Pharma	Industry sector consisting of major multinational pharmaceutical companies.
Biogen	One of the first highly successful biotechnology companies; headquartered in Cambridge, Massachusetts.
Biomarkers	An objective, measurable indicator of a biological state that can be used to measure the presence or progress of disease, or the effects of treatment.
Biotechnology	An industry sector of companies that utilize technology to precisely manipulate biological components to create commercially viable drugs.
BLA	Biologics License Application; application to the FDA for market authorization of a protein-based therapeutic, like an antibody.
Black box warning	Warning that appears on a prescription drug's label in cases where the drug is known to cause serious or life-threatening side effects.
Blinding	The practice whereby both patients and physicians are unaware of the treatment being provided in a clinical trial; done to mitigate bias.

Blockbuster	A drug that generates at least $1 billion in annual revenue.
BMS	Bristol Myers Squibb; pharmaceutical company that developed and commercialized the oncology blockbuster drug Opdivo®.
CDK4/6 pathway	Cellular pathway targeted by breast cancer drugs developed by Pfizer (Ibrance®—first in class), Eli Lilly (Verzenio), and Novartis (Kisqali®).
Celgene	Pharmaceutical company acquired by BMS that earlier purchased Inrebic®, a JAK2 inhibitor to treat myelofibrosis, from the original innovating company, TargeGen.
Chemotherapy	Chemical cancer treatment that typically nonselectively kills rapidly dividing cell, including cancer cells.
CML	Chronic myelocytic leukemia; a leukemia caused by the mutation of a single gene fusion called BCR-*Abl*. Effectively treated by Gleevec®, a Novartis drug.
Conditional approval	An FDA designation also known as accelerated approval that provides marketing authorization prior to the completion of comprehensive, traditional Phase 3 trials.
Contract research organization (CRO)	A business that conducts large-scale clinical drug trials and operates on a fee-for-service plus guaranteed payment model.

GLOSSARY

CTLA-4	Cytotoxic T-lymphocyte-associated protein 4; an immune checkpoint inhibited by Yervoy®. TRACON is currently testing Yervoy® with envafolimab in the ENVASARC trial.
DLBCL	Diffuse large B-cell lymphoma; a cancer of white blood cells called B-lymphocytes. Rituxan® was approved to treat this type of lymphoma in 2006.
Dossier	A comprehensive, auditable collection of documents specific to a drug and its trials required for FDA approval and subsequent commercialization initiatives.
Extrapulmonary tuberculosis	Tuberculosis that infects organs outside of the lungs.
Factor VIII	A key clotting factor that is deficient in hemophiliacs.
FDA	Food and Drug Administration; the US drug review and approval body.
First in class	First-in-class drugs are ones with a new and unique mechanism of action for treating a medical condition.
First line of treatment	Treating a patient as soon as they are diagnosed and prior to treatment with any other agent.
Folotyn®	Drug approved for the treatment of orphan drug indication of peripheral T cell lymphoma that costs more than $30,000 per month. Developed and marketed by Allos Therapeutics.

Genentech	Biopharmaceutical company headquartered in south San Francisco that is part of the Roche conglomerate; markets some of the most commercially successful antibody cancer treatments in the world, including the blockbuster drugs Herceptin®, Avastin®, and Rituxan®.
GIST	Gastrointestinal stromal tumor, a form of sarcoma.
GlaxoSmithKline	Pharmaceutical company that markets Votrient®, the only drug approved to treat refractory UPS and MFS cancers, offering a 4 percent response rate.
Gleevec®	Highly effective drug marketed by Novartis to treat Chronic myelocytic leukemia (CML). Generated multibillion-dollar annual sales treating a relatively small patient population prior to its patent expiration.
Gout	Painful inflammation of the joints, typically the big toe, that develops when uric acid accumulates and crystalizes.
Her2 negative	A type of breast cancer in which cancerous cells do not express the protein Her2.
HIV	Human immunodeficiency virus; a virus that attacks T-lymphocytes, a type of white blood cell that fights infection. Spread by contact with certain bodily fluids of a person infected with HIV, either through unprotected sex or shared needles. Left untreated, HIV can cause AIDS (Acquired Immunodeficiency Syndrome).
Ibrance®	Pfizer's first-in-class blockbuster breast cancer drug.

GLOSSARY

ICI	Immune checkpoint inhibitors; a class of drugs approved to treat more than twenty cancer types. ICIs have revolutionized cancer care by allowing the patient's immune system to attack his or her cancer.
IDEC	Biotechnology firm that discovered and, with Genentech, codeveloped and commercialized Rituxan®, a blockbuster cancer and rheumatology drug.
Incyte	Pharmaceutical company that markets the first-in-class JAK2 inhibitor, Jakafi®, to treat myelofibrosis.
Inrebic®	JAK inhibitor initially developed by TargeGen to treat myelofibrosis. Sold to Sanofi and then Celgene and now is marketed by BMS for the treatment for myelofibrosis.
Intron® A	An antiviral drug marketed by Roche to treat viral infections and cancer. Intron® A was one of the most frequently prescribed treatments for kidney cancer prior to the approval of Sutent®.
JAK2 gene	Mutation of this gene can cause myelofibrosis; drugs that target the JAK2 gene can effectively treat the disease.
Jakafi®	A first-in-class JAK2 inhibitor approved to treat myelofibrosis, marketed by Incyte Pharmaceuticals.
Janssen	The oncology development arm of Johnson & Johnson.
JLABS	Innovation arm of Johnson & Johnson that provides drug-development infrastructure to innovative companies that support discovery efforts.

Johnson & Johnson	American multinational corporation that develops medical devices, pharmaceuticals, and consumer packaged goods; the largest pharmaceutical company in the world.
Keytruda®	Blockbuster cancer drug marketed by Merck, approved for multiple cancer types; predicted by *Barron's* to be the top-selling drug in the world by 2025.
Label expansion	Further testing and approval of a drug after its initial approval to demonstrate its efficacy in patient populations beyond the ones for which the drug was initially approved.
Mechanism of action	How a drug produces a specific pharmacological effect.
Merck	One of the largest pharmaceutical companies in the world; developed and markets Keytruda®, a blockbuster cancer drug predicted to become the top-selling drug in the world by 2025.
Myxofibrosarcoma (MFS)	An uncommon subtype of sarcoma.
Monoclonal antibody (MAB)	Unique antibody produced in the laboratory that binds to a single cell receptor.
NCCN	National Comprehensive Cancer Network; an alliance of thirty cancer centers in the United States dedicated to patient care, research, and education. The NCCN Compendium of cancer drugs informs new treatment uses and supports reimbursement by Medicare and other insurance companies.

GLOSSARY

NDA	New Drug Application; application for market authorization submitted to the FDA for a small-molecule drug.
Novartis	One of the largest pharmaceutical companies in the world. Based in Switzerland. Markets Gleevec®, which treats chronic myelocytic leukemia (CML).
Opdivo®	One of the first approved immune checkpoint inhibitors; developed and marketed by Bristol Myers Squibb.
Orphan drug	A pharmaceutical agent developed to treat a rare medical condition occurring in fewer than two hundred thousand patients in the United States.
Pastan, Ira, MD	Senior investigator at the National Cancer Institute who has made major contributions to the development of ADCs and monoclonal antibodies.
Pfizer	One of the largest pharmaceutical companies in the world. Markets Ibrance® and Sutent®.
Phase 1 Trial	Clinical trial that primarily assesses safety of a drug candidate.
Phase 2 Trial	Clinical trial that primarily assesses early signs of activity of a drug, as well as its safety profile.
Phase 3 Trial	Clinical trial that generally determines activity in comparison to an approved drug to see if the drug candidate is a better treatment.
Refractory	A cancer that becomes resistant to treatment over time.
Registration trial	A definitive or pivotal clinical trial conducted by a drug company to secure FDA approval.

Response rate	A determination of a drug's efficacy based on a measurable decrease in cancer size (e.g., tumor shrinkage). End point used to secure accelerated approval for oncology drugs.
Rituxan®	A revolutionary cancer and rheumatology drug that targets B-cells; remains a standard-of-care treatment for patients with lymphoma and rheumatoid arthritis.
Sanofi	Pharmaceutical company headquartered in France that purchased Inrebic® from TargeGen.
Sarcoma	Broad group of cancers that begin in the bones, muscles, or connective tissues.
Seattle Genetics	A biotechnology company headquartered in Seattle that successfully commercialized multiple ADCs; founded by Dr. Clay Siegall.
Siegall, Clay, PhD	Founder of Seattle Genetics, a biotechnology company that has successfully commercialized multiple ADCs.
Small-molecule drugs	Chemical drugs.
Sugen	Biotechnology company that developed the innovative drug Sutent®. Sugen was purchased by Pharmacia, which was in turn bought by Pfizer in 2003.
Sutent®	Small-molecule drug marketed by Pfizer that inhibits blood vessel and cancer growth that revolutionized the treatment of kidney cancer.
Swanson, Robert	Founder of Genentech.

TargeGen	Biotechnology company that focused research and development efforts on the JAK2 inhibitor drug Inrebic® to treat myelofibrosis.
Uloric®	Gout drug that went off-patent in 2019.
Undifferentiated pleomorphic sarcoma (UPS)	A sarcoma subtype responsive to immune checkpoint inhibition.
Verzenio®	CDK4/6 inhibitor marketed by Eli Lilly for breast cancer patients.
Votrient®	Small-molecule drug marketed by GlaxoSmithKline that inhibits blood vessel and cancer growth; only approved treatment for refractory UPS or MFS.
Yervoy®	Antibody that inhibits the CTLA-4 immune checkpoint marketed by BMS for the treatment of certain cancers.

ABOUT THE LEAD AUTHOR

Dr. Charles Theuer has been CEO and President and a Director of TRACON Pharmaceuticals (NASDAQ: TCON) since 2006. Prior thereto, from October 2004 to July 2006, Dr. Theuer was Chief Medical Officer at TargeGen Inc., where he initiated the development of small molecule kinase inhibitors in oncology (including Inrebic® (fedratinib), a JAK2 inhibitor approved for myelofibrosis), ophthalmology, and cardiovascular disease. From October 2003 to October 2004, Dr. Theuer was Director, Clinical Oncology, at Pfizer, where he led the clinical development of Sutent® (sunitinib); Sutent was approved by the US Food and Drug Administration in January 2006 for treating advanced kidney cancer. Prior thereto, Dr. Theuer held senior positions at IDEC Pharmaceuticals, from June 2002 to October 2003, and the National Cancer Institute developing other agents, including small molecules and monoclonal antibody therapies. Dr. Theuer holds a BS from the Massachusetts Institute of Technology, an MD degree from the University of California, San Francisco, and a PhD from the University of California, Irvine. He completed a residency in general surgery at Harbor-UCLA Medical Center and was Board Certified in general surgery in 1997. Dr. Theuer held academic positions at the National Cancer Institute and at the University of California, Irvine, where he was a member of the Division of Surgical Oncology and Department of Medicine. His previous research involved immunotoxin and cancer vaccine development, translational work in cancer patients, and gastrointestinal cancer epidemiology. He serves as a Board member of TRACON Pharmaceuticals, 4D Molecular Therapeutics (NASDAQ: FDMT), Oncternal Therapeutics (NASDAQ: ONCT), and the nonprofit San Diego Squared, which promotes STEM careers for underrepresented youth. He is married with two adult children and enjoys hiking, coaching basketball, and sculling. His guest columns have appeared in the MIT newspaper *The Tech*.

OUR SERVICES

For more about the TRACON Product Development Platform, please visit traconpharma.com or email us at info@traconpharma.com.